PLASTICS REINFORCEMENT AND INDUSTRIAL APPLICATIONS

PLASTICS REINFORCEMENT AND INDUSTRIAL APPLICATIONS

T.R. CROMPTON

CRC Press
Taylor & Francis Group
Boca Raton London New York

CRC Press is an imprint of the
Taylor & Francis Group, an **informa** business

CRC Press
Taylor & Francis Group
6000 Broken Sound Parkway NW, Suite 300
Boca Raton, FL 33487-2742

First issued in paperback 2019

© 2016 by Taylor & Francis Group, LLC
CRC Press is an imprint of Taylor & Francis Group, an Informa business

No claim to original U.S. Government works

ISBN-13: 978-1-4822-3933-1 (hbk)
ISBN-13: 978-0-367-37746-5 (pbk)

Visit the Taylor & Francis Web site at
http://www.taylorandfrancis.com

and the CRC Press Web site at
http://www.crcpress.com

Contents

Preface

Many mechanical, electrical, thermal, and other properties, such as thermal stability, of plastics can be considerably improved by the incorporation of reinforcing agents, such as glass fiber, carbon fiber, carbon nanotubes, and many more fillers, such as talc and clay in their formulation.

This makes it possible, for example, to produce plastics that are suitable for use in the construction of aircraft and lightweight automobiles and for use in high-temperature applications in aircraft and electro- and automobiles.

This book is a thorough review of developments in this field, and it is hoped it will familiarize the reader with developments in this very exciting area.

Author

T.R. Crompton is the former head of plastics research in analytical methods at Shell Chemicals and is currently the head of the water testing department of a leading water authority. He has published approximately 50 books in these fields.

1 Introduction

Many of the mechanical and other properties of plastics can be considerably improved by the incorporation of reinforcing agents in their formation. The incorporation of fillers such as clay or talc in plastics formulation can also have beneficial effects on the properties of plastics.

Typical reinforcing agents now commonly employed in the formation of plastics include glass fiber, carbon fiber, carbon nanotubes, carbon black, graphite, organically modified clay, talc, graphene, and fullerene.

With correct formulation and blending, the tensile and impact strength and modulus of elasticity can be improved by factors of 3 or more, enabling plastics to be used in applications previously undreamt of.

Many other polymer properties, such as stress, compressive set, strain, shear strength, friction coefficient, shrinkage, toughness, abrasion resistance, stress relation compressive creep, weatherability, and resistance to various types of radiation, can be of importance in polymers, whether or not they are reinforced. Methods for measuring these properties are discussed in Chapter 2.

Typical values for these parameters are compared in Chapter 3 for reinforced polymer plastics and, where relevant, for the unreinforced counterpart.

This survey covers measurements on plastics with all the reinforcing agents previously mentioned and includes a wide range of plastics now being used in plastics technology. It includes commonly used plastics such as polyamides, polyesters, polyethylene terephthalate, and epoxy resins, but also covers newer plastics, such as polyimides, polysulfones, polyethersulfone, polyphenylene sulfide, and polyether ether ketone, all of which have more specialized applications.

As discussed in Chapter 4, in addition to the mechanical properties, incorporation of reinforcing agents or fillers can affect the thermal properties of plastics, such as heat resistance.

This is used in polymers intended for high-heat applications, such as expansion tanks, microwave ovens, and electrical connectors.

The thermal expansion of plastics can be increased by the incorporation of reinforcing agents or fillers into the formulation. Thus, the incorporation of ceramic powder filler into polytetrafluroethylene (PTFE) reduces the coefficient of thermal expansion. On the other hand, the incorporation of 20% glass fiber into epoxy resins will increase the coefficient of cubical expansion from 0.5 mn/mn/°C $\times$ 10^{-5} to 2.0 mn/mn/°C $\times$ 10^{-5}. Reinforcement of perfluoroalkoxyethylene improved the heat distortion temperature at 0.45 MPa from 24°C to 100°C and at 1.8 MPa from 30°C to 58°C. This was accompanied by a nominal increase in tensile strength.

A very important recent development is the production of plastics such as carbon fiber–reinforced polyimides. These have, for example, very distinct applications in aerospace jet engines, which retain their mechanical and other properties at temperatures of 350°C or higher. Developments such as this one are proceeding rapidly.

Glass-reinforced polyphenylene sulfide is another plastic with a high heat resistance. It can withstand temperatures of up to 180°C without changes in properties and, as such, has found applications in electronics, food packaging, and transportation.

As discussed in Chapter 5, unreinforced plastics have been used in electrical applications as diverse as wire and cable covering, insulation, transformers, switch gear, electric motor components, capacitors, printed boards, and electronic compounds. Reinforcement of the polymer with 20%–40% glass fiber, for example, extends the range of applications to automobile ignition switches, relay switches, terminal boards, electrical and electronic devices, encapsulation, electrical circuit breakers, brush holders in electric motors, termination blocks, and electrical plugs.

Surface resistivity is one factor that can be controlled by the incorporation of a reinforcing agent or a filler into a polymer formulation. Thus, the incorporation of natural fiber into high-density polyethylene composites decreases the surface resistivity of the composite. Reinforcing agents can also affect the electrical resistance. The incorporation of Kevlar fibers in polyaniline increases the tracking resistance threefold over the value for the base polymer.

The dissipation factor is another electrical property that can be affected by the incorporation of reinforcing agents or fillers into the formulation of a plastic. Thus, the incorporation of 10%–60% glass fiber into polycarbonate and polyamide 6,6 considerably improves the dissipation factor.

Virgin polyether ether ketone has excellent electrical and mechanical properties. However, the mechanical properties can be further improved by the incorporation of 30% glass fiber without compromising the electrical properties. Thus, the glass fiber improves the tensile strength of polyether ether ketone from 92 to 151 MPa, with a severe drop in the elongation of break. Carbon fiber reinforcement causes a deterioration in the volume resistivity of epoxy resins and polytetrafluoroethylene, but improves the dielectric strength and dielectric constants, with a deterioration in the surface resistance and tracking resistance. The incorporation of 30% carbon fiber into polyoxymethylene causes a deterioration in volume resistivity and dielectric strength while having very beneficial improvements in tensile strength, flexural modulus, heat distortion temperature, and maximum operating temperature.

Enough has been said to illustrate the fact that a careful study of the effect on mechanical, thermal, and electrical properties of a plastic often makes it possible to choose plastic formulations that will meet the requirements for a particular application of plastics.

Nanotechnology is deeply embedded in advanced devices for electronic and optoelectronic applications.

The utility of polymer-based nanocomposites in these areas is quite diverse, involving many potential applications, as well as types of nanocomposites.

One specific nanocomposite type receiving considerable attention involves conjugated polymers and carbon nanotubes. There is a litany and growing application of these involving electronically conducting polymers, photovoltaic cells, light-emitting diodes, and field effect transitions.

The electrical conductivity of carbon nanotubes in insulating plastics is a topic of considerable interest. The potential applications include electromagnetic interference shielding, transparent conductive coatings, electrostatic dissipation, and supercapacity actuators. Thus, the threshold conductivity of single-walled carbon nanotube–epoxy resins was noted to be a function of the type of single-walled carbon nanotubes used in the formulations.

The threshold conductivity of epoxy nanocomposites had values as low as 0.00005% volume traction of carbon black fraction, compared to a 1.5 volume fraction in acrylic emulsion films.

Electronically conducting polymers already have a wide range of applications, and it is entirely possible that we may someday see conducting plastic replacing copper or aluminum cable for power transmission.

Carbon nanotubes or exfoliated graphite, that is, graphene, offers substantial opportunities in the electrical and electronic industries as well as potential specific emerging techniques. Considerable work is being carried out in this area testing graphene and also fullerene, and this will, undoubtedly, produce new, very exciting materials.

As discussed in Chapter 6, the incorporation of reinforcing agents or fillers into plastic formulations can, in some but not all cases, lead to variations in the molecular stability of plastics and also their thermal and thermooxidation stability. Thus, it has been observed that the addition of silica to polytetrafluoroethylene did not adversely affect polymer stability, while the incorporation of 25% of organically modified silica into polyethylene led to a decrease in weight loss of the plastic from 80% to 33.7%. The incorporation of carbon nanotubes in epoxy resins improved their mechanical and thermal properties. It is fair to say that the effect of reinforcing agents on the thermal and thermooxidative stability of polymers must always be born in mind when selecting polymer formulations for a particular application.

Reinforced plastics are used in a wide range of industries, including general engineering; automotive, aerospace, and space vehicles; electronics; microwaves; computers; food, beverage, and drug packaging and preservation; medical devices; nuclear; and may others.

The use of plastics in engineering applications is growing exponentially and has been adopted by a wide range of manufacturers to improve the properties of their product and to produce what are new fabrication materials with previously unattainable properties.

A classic case, perhaps, is the development of the Boeing 787 Dreamliner aircraft, in which preimpregnated carbon fiber is used in an epoxy matrix composite and in toughened polyacrylonitrile-based fibers. This aircraft contains about 50% by weight of plastic components, including fuselage, wings, landing gear, horizon stabilizers, passenger doors, wing tips, ice protection system, windows, tail assembly, and hydraulic clamps.

A further example of frontline development is the use of reinforced polyimides for the construction of jet engine nozzles that will withstand severe temperatures of several hundred degrees centigrade. Reinforced plastics are now being used in a wide variety of industries, ranging from the nuclear industry, to medical instruments and implants, to food packaging where polymers are exposed to gamma radiation, down to many basic applications, such as gears.

The future for reinforced plastics is assessed, and it is hoped that this book will prove invaluable to those working in this very exciting field.

2 Measurement of Mechanical Properties of Reinforced Plastics

2.1 INTRODUCTION

In the past, results of standard tests such as tensile strength, Izod impact strength, and softening point have been major emphases in the technical literature on plastics. More recently, however, with the increasing use of plastics in more critical applications, there has been a growing awareness of the need to supplement such information with data obtained from tests more closely simulating operational conditions.

For many applications, the choice of material used depends upon the balance of stiffness, toughness, processability, and price. For a particular application, a compromise between these features will usually be necessary. For example, it is generally true that within a given family of grades of a particular polymer, the rigidity increases as impact strength decreases. Again, processing requirements may place an upper or a lower limit on the molecular weight of the polymer that can be used, and this will frequently influence the mechanical properties quite markedly. Single test values of a given property are a less reliable guide to operational behavior with plastics than with metals. This is because for plastics, the great mass of empirical experience and tradition of effective design that has been built up over centuries in the field of metals is not yet available. Furthermore, no single value can be placed on the stiffness or the toughness of a plastic because

- Stiffness will vary with time, stress, and temperature.
- Toughness is influenced by the design and size of the component, the design of the mold, processing conditions, and the temperature of use.
- Stiffness and toughness can be affected by environmental effects, such as thermal and oxidative aging, ultraviolet (UV) aging, and chemical attack (including the special case of environmental stress corrosion).

In addition, a change in a specific polymer parameter may affect processability and basic physical properties. Both of these factors can interact in governing the behavior of a fabricated article. Comprehensive experimental data are therefore necessary to understand effectively the behavior of plastic materials, and to give a realistic and reliable guide to the selection of material and grade.

In many applications, plastics are replacing traditional materials. Hence, there is often a natural tendency to apply to plastics tests similar to those that have been found suitable for gauging the performance of the traditional material. Dangers can obviously arise if plastics are selected on the basis of these tests without clearly recognizing that the correlation between values of laboratory performance and field performance may be quite different for the two classes of materials.

The mechanical properties of polymers can be considerably improved by the incorporation in their formulation of reinforcing agents or fillers. These have been used to improve or alter the mechanical properties of polymers. These include glass fiber, glass beads, calcium carbonate, minerals, mica, talc, clay, carbon fiber, carbon nanotubes, aluminum or other metal powders, silica and silicones, and others [1–16].

This is the subject matter of this book. It is believed that the incorporation of such agents in polymer formulations will play an important part in the development of polymers with improved properties, which will play a very important part in the development of applications of plastic material in engineering.

Some of the important mechanical properties used to evaluate unreinforced polymers are listed in Table 2.1 for a range of virgin plastics. These methods can also be used to evaluate reinforced plastics, as will be seen later.

2.2　TENSILE STRENGTH ELECTRONIC DYNAMOMETER

ATS FAAR supplies the Series TC200 computer-governed dynamometer, which can carry out tensile, compression, and flexural tests on a variety of material. All details of the tests are managed, including computing the final results and presenting them in alphanumerical or graphical form.

Tests that can be carried out by the dynamometer include those listed below and those shown in Table 2.2:

- Tensile strength (tensile modulus) tests with or without preloading, according to specifications ASTM D638-03 [17], DIN EN ISO 527-1 [18], and DIN EN ISO 527-2 [19]
- Compression and compressive strength tests according to specifications ASTM D695-02a [20] and DIN EN ISO 179-1 [21]
- Flexural and flexural strength tests according to specifications ASTM D790-03 [22], ASTM D732 [23], and DIN EN ISO 178 [24]

Polymers of excellent tensile strength (i.e., >100 MPa) include epoxies, polyester laminates, polyamide (PA) 4/6, polyamide-imide, and polyvinylidene fluoride.

2.3　FLEXURAL MODULUS (MODULUS OF ELASTICITY)

This is the short-term modulus of materials at specified temperatures. Ratings for flexural modulus have been assigned at 20°C and are usually determined at ~1% strain. An *excellent* rating indicates a high flexural modulus; a *very poor* rating indicates a low flexural modulus. A *not applicable* status indicates that the material has a modulus of limited practical use.

TABLE 2.1
Mechanical Properties of Polymers

Polymer	Tensile Strength (MPa) ASTM D638[a]	Flexural Modulus (Modulus of Elasticity) (GPa)	Elongation at Break (%) ASTM D638	Strain at Yield (%)	Notched Izod Impact Strength (kJ/m)	Surface Hardness	Compression Strength ASTM D695 (MN/M²)
		Carbon/Hydrogen-Containing Polymers					
Low-density polyethylene (LDPE)	10	0.25	400	19	1.064	SD 48	
High-density polyethylene (HDPE)	32	1.25	150	15	0.15	SD 68	16.5
Cross-linked polyethylene (PE)	18	0.5	350	N/Y	1.064	SD 58	
Polypropylene $CaCO_3$ (PP)	26	2	60	N/Y	0.05	RR 85	59–69
Ethylene-propylene	26	0.6	500	N/Y	0.15	RR 75	
Polymethylpentene	28	1.5	15	6	0.04	RR 70	
Styrene-butadiene	28	1.6	50	N/Y	0.08	SD 75	
Styrene-butylene-styrene	6	0.02	800	N/Y	1.06	SA 45	
High-impact polystyrene (PS)	42	2.1	2.5	1.8	0.1	RM 30	27–62
PS, general purpose	34	3	1.6				79–110
		Oxygen-Containing Polymers					
Epoxies general purpose	60–80	3–3.5	4–8	N/A	0.5	RM 113	
Acetal (polyoxymethylene)	50	50	27	20	8	0.10	RM 109
Polyesters (bisphenol), polyester laminate (glass filled)	280	16	1.5	N/A	1.064	RM 125	
Polyester (electrical grade)	40	9	2	N/A	0.4	RM 125	
Polybutylene terephthalate	52	2.1	250	4	0.06	RM 70	
Polyethylene terephthalate (PET)	55	2.3	300	3.5	0.02	RM 30	
Polyether ether ketone (PEEK)	92	3.7	50	4.3	0.083	RM 99	
Diallyl isophthalate	82	11.3	0.9	N/A	0.37	RM 112	
Diallyl phthalate	70	10.6	0.9	N/A	0.41	RM 112	
Alkyd resin glass fiber reinforced	72	8.6	0.8	N/A	0.24	RM 125	

Continued

TABLE 2.1

Mechanical Properties of Polymers

Polymer	Tensile Strength (MPa) ASTM D638[a]	Flexural Modulus (Modulus of Elasticity) (GPa)	Elongation at Break (%) ASTM D638	Strain at Yield (%)	Notched Izod Impact Strength (kj/m)	Surface Hardness	Compression Strength ASTM D695 (MN/M²)
Polyarylates	68	2.2	50	8.8	0.29	RR 125	
Polycarbonate (PC)	50	2.1	200	3.5	0.05	RM 70	
Polyphenylene oxide	65	2.5	60	4.5	0.16	RR 119	
Phenol formaldehyde	45	6.5	1.2	N/A	0.024	RM 114	
Styrene-maleic anhydride	52	3	1.8	2	0.03	RL 105	
Cellulose acetate	30	1.7	60	4	0.26	RR 71	
Cellulose propionate	35	1.76	60	4	0.13	RR 94	
Cellulose butyrate acrylics	70	2.9	2.5	N/A	0.02	RM 92	
Ethylene vinyl acetate	17	0.02	750	N/A	1.66	SA 85	
Nitrogen-Containing Polymers							
Polyamide (PA) 6	40	1	60	4.5	0.25	SD 75	
PA 4,6	100	1	30	11	0.1	SD 85	
PA 11	52	0.9	320	20	0.05	RR 105	
PA 6,9	50	1.4	15	10	0.06	SD 78	
PA 12	50	1.4	200	6	0.06	RR 105	
PA 6,6	59	1.2	60	4.5	0.11	RR 90	
PA 6,12	51	1.4	300	7	0.04	RR 105	
Nylon/acrylonitrile (ABS) alloy	47	2.14	270	6	0.85	RR 99	
PA-imide	185	4.58	12	8	0.13	RM 109	
Polyimide	72	2.45	8	4	0.08	RM 100	
Polyetherimide	105	3.3	60	8	0.1	RM 109	
Polyurethane thermoplastic elastomer	24	0.003	700	N/Y	1.064	SA 70	

Material						
Ether ester amide elastomer	57	10	0.6	N/A	0.02	RM 115
Styrene-acrylonitrile	72	3.6	2.4	3.5	0.02	RM 80
ABS	34	2.1	6	2	0.18	RR 96
Acrylate-styrene-acrylonitrile	35	2.5	10	3.3	0.1	RR 106
Fluorine-Containing Polymers						
Polytetrafluoroethylene	25	0.70	400	70	0.16	RM 69
Polyvinyl fluoride	40	1.4	150	30	0.18	SD 80
Polyvinylidene fluoride	100	5.5	6	N/A	0.12	SD 90
Perfluoroalkoxyethylene	29	0.7	300	85	1.064+	SD 60
Ethylene tetrafluoroethylene	28	1.4	150	15	1.064+	RR 50
Ethylene chlorotrifluoroethylene	30	1.7	200	5	1.064+	RR 93
Fluorinated ethylene-propylene	14	0.6	150	6	1.064+	RR 45
Chlorine-Containing Polymers						
Chlorinated polyvinyl chloride (PVC)	58	3.1	30	5	0.06	SA 70
Unplasticized PVC (UPVC)	51	3	60	3.5	0.08	RR 110
Plasticized PVC	14–20	0.0007–0.03	280–295	N/Y	1.05	SA 85
Sulfur-Containing Polymers						
Polyphenylene sulfide	91	13.8	0.6	N/A	0.6	RR 121
Polysulfone	70	2.65	80	5.5	0.07	RM 69
Polyethersulfone	84	2.6	60	6.6	0.084	RM 85
Silicon-Containing Polymers						
Silicones	28	3.5	2	N/A	0.02	RM 80

a These are typical room temperature values of notched Izod impact strength. A material that does not break in the Izod test is given a value of 1.06+ kJ/m; the + indicates that it has a higher impact energy than the test can generate.

N/A = not applicable, if material is brittle and does not exhibit yield point; N/Y = no yield point, if material is ductile and does not exhibit yield point; RM = Rockwell M 123 hardness (hard); RR = Rockwell R 112 hardness; SA 65 = Shore A 65 hardness (soft); SD 75 = Shore D75 hardness.

TABLE 2.2
Mechanical Properties of Polymers Using the Dyanometer Test

Method	AST FAAR Apparatus Code Number	Measurement Units	Test Suitable for Meeting the Following Standards
Tensile impact	16.10050	kj/m^2	DIN EN ISO 8256 [25]
Tensile strength (tensile modulus)	16.00121	N/mm^2	ASTM D638-03 [17]
	16.00122	psi	DIN EN ISO 527-1 [18]
	16.00126	kg/cm^2	DIN EN ISO 527-2 [19]
	16.00127		
Compression strength and compressive	16.00121	N/mm^2	ASTM D695-02a [20]
stress	16.00122	psi	DIN EN ISO 604 [26]
	16.00126	N/mm^2	
Flexural strength (Dyanometer)	16.00121	N/mm^2	ASTM D790-03[27]
	16.00122	psi	ASTM D732 [28]
	16.00126	kg/cm^2	DIN EN ISO 178 [24]
	16.00127		

ASTM = American Society of Testing and Materials, DIN EN ISO = Deutsches Institut für Normung Europa Norm International Standards Organization.

2.3.1 TORSION TEST

AST FAAR supplies a torsion test instrument for the measurement of the apparent modulus of elasticity versus temperature of plastics and elastomers (Table 2.3) following specifications ASTM D1043 [29], DIN 53477 [30], ISO 498 [39], and BS 2782 [32].

It is possible to measure accurately the apparent modulus of elasticity of specimens obtained from a wide range of materials. The modulus, in this case, is defined as apparent because it is obtained by measuring the angular torsion of the specimens under test.

This apparatus can conduct tests in the temperature range −100°C to 300°C.

In Table 2.4 are listed those polymers in their virgin (i.e., unreinforced) state that have a high tensile strength and those that have a high flexural modulus. Further improvements in these properties can be achieved by incorporation of reinforcing agents, which will be discussed later.

In selecting a polymer for a particular end-use application, it is frequently necessary to compromise on polymer properties. Thus, while epoxy resins have excellent tensile strength and flexible modulus, as well as detergent resistance, they have low resistance to gamma radiation, poor heat distortion performance, and poor wear properties. They are also expensive and have a poor volume sensitivity and surface finish.

2.3.2 HAND TEST

ATS FAAR also supplies hand test apparatus for the measurement of elasticity (flexural modulus) according to specifications ASTM D1869 [33], ASTM D3419 [34],

TABLE 2.3
Modulus of Elasticity of Polymers

Method	ATS FAAR Apparatus Code Number	Measurement Units	Test Suitable for Meeting the Following Standards
Modulus elasticity versus temperature (torsion test)	10.22000 10.22001	GPa	ASTM D1043 [29] DIN 53477 [30] ISO 458 [31] BS 2782 [32]
Modulus of elasticity or flexural modulus (bend test)	10.2900	kN/mm^2 psi kg/cm^2	ASTM D1986 [33] ASTM D3419 [34] ASTM D3641 [35] ASTM D4703 [36] ASTM D5227 [37, 38] DIN EN ISO 178 [24] DIN EN ISO 527-1 [18] DIN EN ISO 527-2 [19] DIN EN ISO 604 [26]

ASTM D3641 [35], ASTM D4703 [36], ASTM 5227 [37], DIN EN ISO 178 [24], DIN EN ISO 527-1 [8], and DIN EN ISO 604 [26] (see Table 2.3).

Janick and Kroliskowski [40] investigated the effect of Charpy notched impact strength on the flexural modulus of polyethylene and polyethylene terephthalate. Polymers with good flexural modulus include polydiallylisophthalate (11.3 GPa), phenol-formaldehyde (6.5 GPa), alkyd resins (8.6 GPa), and polyphenylene sulfide (13.8 GPa), as well as glass-filled polyester laminate (16 GPa) epoxy resins (3–3.5 GPa), silica-filled epoxies (15 GPa), and acetals containing 30% carbon fiber (17.2 GPa).

2.4   ELONGATION AT BREAK

Elongation at break is the strain at which the polymer breaks when tested in tension at a controlled temperature (i.e., the tensile elongation at specimen break).

An increasing number of applications are being developed for thermoplastics in which a fabricated article is subjected to a prolonged continuous stress. Typical examples are pipes, crates, cold-water tanks, and engine cooling fans. Under such conditions of constant stress, materials exhibit (to varying extents) continuous deformation with increasing time. This phenomenon is called creep.

A wide variety of materials will, under appropriate conditions of stress and temperature, exhibit a characteristic type of creep behavior (Figure 2.1).

The general form of this creep curve can be described as follows. Upon application of the load, an instantaneous elastic deformation occurs (O–A). This is followed by an increase in deformation with time as represented by the portion of

TABLE 2.4

Identification of Reinforced Polymers with Outstanding Tensile Strength and Flexural Modulus

Polymer	Tensile Strength (MPa)	Flexural Modulus (GPa)	Elongation at Break (%)	Strain at Yield (%)	Notched Izod (kj/m)	Excellent or Good Performance in the Following	Poor Performance in the Following
Epoxies	60–80	3–3.5	4–8	N/A	0.5	Tensile strength, flexural modulus, detergent resistance, resistance to gamma radiation, heat distortion temperature (HDT), wear properties, low resin shrinkage during cure	Poor surface finish, high cost, elongation at break, volume resistivity
Polyesters (bisphenol polyester laminate), glass filled	280	16	1.5	N/A	1.06+	Impact strength and cheap when compared with epoxies	Heat resistance, solvent resistance, elongational at break
Polyester, sheet molding compound	70	11	2.5	N/A	0.80	High flexural modulus heat distortion temperature (HDT), resistance to UV and gamma irradiation, toughness	More expensive than polyesters, dielectric constant, dielectric strength, flame spread, hydrolytic stability, elongation stability, HDT
PA 4,6	100	1	30	11	0.1	Reasonable HDT, good chemical resistance	Moisture absorption
Polyvinylidene fluoride (20% carbon fiber reinforced)	100	5.5	6	N/A	0.12	Tensile strength, flexural modulus, HDT, detergent resistance, hydrolytic stability	Elongation at break, expensive, gamma irradiation resistance, dielectric properties, surface finish, toughness

						Advantages	Disadvantages
Polyetherimide	100	3.3	60	8	0.10	Tensile strength, cost, oxygen index, detergent resistance, gamma irradiation and UV resistance	Stress cracking with chlorinated solvents, notched impact strength, high cost, tracking resistance, flexural modulus toughness
Polyphenylene sulfide (glass fiber reinforced)	91	13.8	0.6	N/A	0.031	Flexural strength, modulus	Electrical properties, notched Izod
Diallyl isophthalate	82	11.3	0.9	N/A	0.37	Tensile strength, flexural modulus, notched Izod, detergent resistance, HDT, tracking resistance, UV radiation	Volume resistance, dielectric strength, dissipation factor, flammability, hydrolytic stability, high cost, elongation at break
Diallyl phthalate	70	10.6	0.9		0.41	Tensile strength, flexural modulus, notched Izod, detergent resistance, HDT, tracking resistance, UV radiation	Volume resistance, dielectric strength, dissipation factor, flammability, hydrolytic stability, high cost, elongation at break
Polysulfone	70	2.65	80	5.5	0.07	High-temperature performance, electrical properties, hydrolytic resistance, tensile strength, dimensional stability	Surface finish
Polyacrylates	68	2.2	50	15	1.06+[a]	Impact strength, heat resistance, UV resistance	Stress cracking with hydrocarbons, need 350°C processing temperature

[a] These are typical room temperature values of notched Izod impact strength. A material that does not break in the Izod test is given a value of 1.06+ kJ/m; the + indicates that it has a higher impact energy than the test can generate.
N/A = not applicable.

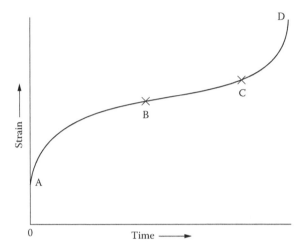

FIGURE 2.1   Creep behavior of PE.

the curve A–D. This is the generally accepted classic curve, and it is usually divided into three parts:

- A–B: The primary stage in which the creep rate decreases linearly with time
- B–C: The secondary stage, where the change in dimensions with time is constant (i.e., a constant creep rate)
- C–D: The tertiary stage, where the creep rate increases again until rupture occurs

With thermoplastics, the secondary stage is often only a point of inflection, and the final tertiary stage is usually accompanied by crazing or cracking of the specimen or, at higher stresses, by the onset of necking (i.e., marked local reduction in the cross-sectional area). In assessing the practical suitability of plastics, we are interested in the earlier portions of the curve, prior to the onset of the tertiary stage, as well as in the rupture behavior.

As the values of the stress, temperature, and time of loading vary with differing applications, the type of information extracted from these basic creep data will also vary with the application. There are several alternative methods (some of which are discussed later in this book) of expressing this derived information, but whichever form is used (whether tabular of graphical), it is unlikely to cover all eventualities. It is thus considered of most value to give the actual creep curves and to discuss by means of examples the methods by which particular data can be extracted from these basic creep curves.

Typical curves for polyethylene and polypropylene at 23°C are shown in Figures 2.2 and 2.3. Lin and coworkers [41] discussed the accuracy of creep phenomena in reinforced PA and polycarbonate (PC) composites. In this paper, the phenomenon in increasing dynamic creep and temperature under tension–tension fatigue loading is compared between semicrystalline and amorphous composites.

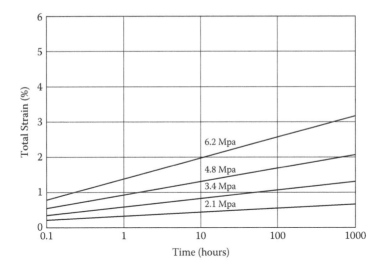

FIGURE 2.2   Creep of high-density polyethylene (HDPE) at 23°C.

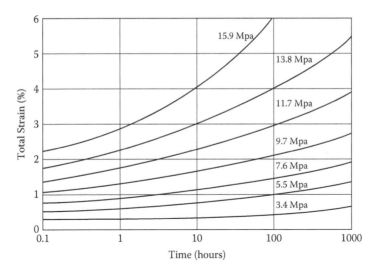

FIGURE 2.3   Creep of PP homopolymer at 23°C.

As shown in Table 2.1, polymers with outstanding elongation at break include low-density polyethylene (400%), styrene-ethylene-butylene-styrene (800%), polyethylene terephthalate (300%), ethylene-vinyl acetate (750%), vinyl acetate (750%), polyurethanes (700%), polyamide (200%), polyamide 612 (300%), and polyamide II (320%).

High elongation at break is not necessarily accompanied by high tensile strength.

Polyethylene terephthalate and various polyamides combine excellent tensile strengths and elongation at break.

Janick and Kroliskowski [42] studied the effect of properties such as elongation at break, tensile strength, flexural strength, elastic modulus, and flexural modulus on the elongation at break performance of polyethylene terephthalate, low-density polyethylene blends, and polyethylene terephthalate polypropylene blends.

2.5   STRAIN AT YIELD

This is an indication of the degree of strain that a material can accept without yielding. The values given are typical room temperature values. If a material is brittle and does not exhibit a yield point, then the value given in Table 2.1 is termed N/A (i.e., not applicable).

   If the material is ductile and does not exhibit a yield point, then the polymer is classified as N/Y (i.e., no yield point).

2.5.1   Isochronous Stress–Strain Curves

Where it is necessary to compare several different materials, basic creep curves alone are not completely satisfactory. This is particularly so if the stress levels used are not the same for each material. If the stress endurance time relevant to a particular application can be agreed on, a much simpler comparison of materials for a specific application can be made by means of isochronous stress–strain curves.

   As an example, let us consider plastic molding in which we can assume that the molding will not be continuously loaded for times >100 h. From the basic creep curves, the strain values at the 100 h point for various stresses can be readily determined. For each material, a stress–strain curve can now be drawn corresponding to the selected loading time of 100 h. Now let us suppose that for this application it is further stipulated that after 100 h of continuous loading, the strain shall be ≤1%. The stress that can be sustained without violating the strain-endurance stipulation can then be conveniently read off for each material from the isochronous stress–strain curves, thereby indicating with reasonable confidence the stress that can be used for design purposes for each of the materials considered. Examples of these interpolations are demonstrated in Figure 2.4.

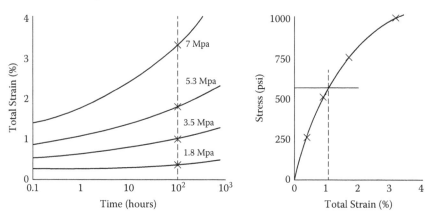

FIGURE 2.4   Isochronous stress–strain curve of PE.

Hay and coworkers [43] developed an approximate method for the theoretical treatment of pressure and viscous heating effects on the flow of a power-law fluid through a slit die. The flow was assumed to remain one-dimensional, and the accuracy of this approximation was checked via finite element simulations of the complete momentum and energy equations. For pressures typically achieved in the laboratory, it was seen that the one-dimensional approximation compared well with the simulations. The model therefore offered a method of including pressure and viscous heating effect in the analysis of experiments, and was used to ationalizes experimentally pressure profiles obtained for the flow of polymer melts through a slit die. Data for the flow of low-density polyethylene and polystyrene melt slit die showed that these two effects were significant under normal laboratory conditions. The shear stress–strain rate curves would thus be affected to the point of being inaccurate at high shear rates. In addition, it was found that the typical technique to correct for a pressure-dependent viscosity was also inaccurate, being affected by the viscous heating and heat transfer from the melt to the die.

Thermomechanical analysis is an ideal technique for analyzing fibers because the measured parameters—dimension change, temperature, and stress—are major variables that affect fiber processing. Figure 2.5 shows the thermal stress analysis curves for a polyolefin fiber as received and after cold drawing. In this experiment, fibers are subjected to initial strain (1% of initial length), and the force required to maintain that fiber length is monitored. As the fiber tries to shrink, more force must be exerted to maintain a constant length. The result is direct measurement of the shrink force of the fiber. Shrink force reflects the orientation frozen into the fiber during processing, which is primarily related to the amorphous portions of the fiber. Techniques that track fiber crystallinity are therefore not as sensitive a measure of processing conditions as thermomechanical analysis. In this case, the onset of the peak of the shrink force indicates the draw temperature, whereas the magnitude of

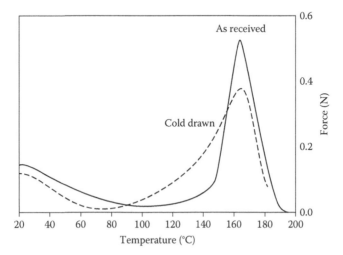

FIGURE 2.5   Application of thermomechanical analysis to thermal stress analysis of polyolefin films (as received and cold drawn).

the peak is related to the draw ratio of the fiber. It has been shown that the area under the shrink force curve (from the onset to maxima) can be correlated to properties such as elongation at break and knot strength. Other portions of the thermomechanical analysis thermal stress plot can yield additional information. For example, the initial decreasing slope is related to the expansion properties of the fiber, and the appearance of secondary force peaks can be used to determine values such as heat set temperature in polyamide.

2.5.2 STRESS–TIME CURVES

With thermoplastic materials, above a certain stress, the stress–strain curve shows a marked departure from linearity; that is, above such values, further increases in stress lead to disproportionately greater increases in elongation. By study of the isochronous stress–train curves of a material, it is possible to decide upon a certain strain that should not be exceeded in a given application. The stress required to produce this critical strain will vary with the time of application of the load, so that the longer the time of application, the lower will be the permissible stress.

From isochronous stress–strain curves relating to endurance times of, for example, 1, 10, 100, and 1,000 h, the magnitude of stress to give the critical strain at each duration of loading can be readily deduced. This procedure can be repeated for different selected levels of strain. In general, the more critical the application and the longer the time factor involved, the lower will be the maximum permissible strain. For the polyolefin family of materials, the upper limit of critical strain will always be governed by the onset of a brittle-type rupture (e.g., hair cracking at elongations that are low compared with those expected from the short-term ductility of the material).

With thermoplastics of intermediate modules (e.g., high-elasticity polyethylene, polypropylene), the stress–strain curves depart slightly from linearity at quite low strains. Thus, if accurate results are to be obtained from standard formulae, it is sometimes necessary to limit the critical strain and the corresponding design stress to low values. One commonly selected criterion with high-modulus thermoplastics (e.g., polyacetals) is to base the design stress and the design modulus on that point on the stress–strain curve at which the secant modulus falls to 85% of the initial tangent modulus. This procedure is illustrated in Figure 2.6.

2.5.3 STRESS–TEMPERATURE CURVES

The methods discussed so far have been concerned with the presentation of combined creep data obtained at a single temperature. Thus, a further method is required to indicate the influence of temperature.

One convenient method is to combine the information available from isochronous stress–strain curves or stress–time curves obtained on the same materials at different temperatures. For example, suppose the performance criterion for a particular application is that the total strain should be ≤2% in 1,000 h. Using the 1,000 h isochronous stress–strain curve for each temperature and erecting an ordinate at the 2% point on the strain axis, the individual working stresses for each temperature can be obtained. Alternatively, by erecting an ordinate at the 1,000 h point on the stress–time curve

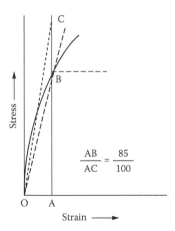

FIGURE 2.6   Stress–time curve.

for 2% strain for each temperature investigated, the individual working stress can be similarly obtained. From these interpolated results, the stress–temperature curve can be drawn down.

By a similar procedure, stress–temperature plots for other specific time-permissible strain combinations can be obtained.

2.5.4   RUPTURE DATA

In applications in which strains up to the order of a few percent do not interfere with the serviceability of the component, the controlling factor as far as permissible stress is concerned will probably be the rupture behavior of the material. The rupture curve forms the limiting envelope of any family of creep curves. Care must therefore be exercised in extrapolating creep date to ensure that for long-term applications, a suitable safety factor is always allowed over the corresponding extrapolated rupture stress. In general, with thermoplastics, an increase in molecular weight leads to improved rupture resistance, which is not necessarily reflected by an improvement in creep resistance, and in fact may be associated with a decrease in creep resistance. An increase in molecular weight will also, in general, be associated with a decrease in processability, which will tend to give higher residual stresses in fabricated articles, which can in turn influence rupture properties. In designing injection-molded articles, particular care should be taken to ensure that the principal stresses experienced by the article in service are (wherever possible) parallel to the direction of flow of material in the mold. Avoidance of stress concentration effects, by appropriate design, will tend to minimize the interference of rupture phenomena and allow greater freedom to design on the basis of the creep modulus.

Certain environments reduce rupture performance with particular materials (e.g., polyethylene in contact with detergents), and allowance must be made for this, where necessary, by the incorporation of an additional safety factor to the rupture stress. In general, the resistance to rupture will be greater if the stress is applied in compression than in tension.

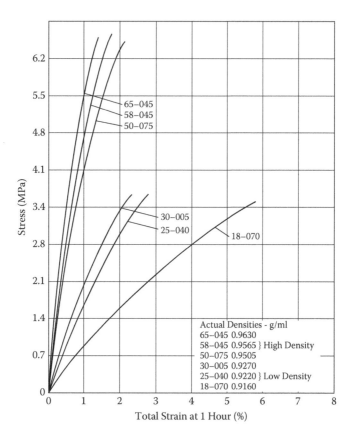

FIGURE 2.7 Isochronous stress–strain curves of various grades of LDPE and HDPE at 23°C (1 h data).

Differing stress levels have been used for high-density polyethylene and low-density polyethylene for the reasons previously discussed, so an easier comparison between grades can be made by the use of isochronous stress–strain curves.

These are shown in Figures 2.7 and 2.8 for times of 1 and 1,000 h, respectively.

The data in Figure 2.9 demonstrate the marked influence of density on the creep behavior of polyethylene. The curves in Figure 2.9 are relevant to a total strain of 1%, but similar plots for other permissible strains can be readily derived from the isochronous stress–strain curves. The linear relationship between creep and density for polyethylene at room temperature, irrespective of the melt index over the range investigated (i.e., 0.2–5.5), has enabled the stress–time curve of Figure 2.10 to be interpolated for the complete range of polyethylene. In this case, the data have been based on a permissible strain of 2%, but as previously explained, data for other permissible strains can be similarly interpolated from the creep curves.

For polyethylene, the principal factor controlling creep is density, but the principal factor influencing rupture behavior is the melt index. Thus, for a given density, for long-term applications, the higher the melt index, the lower will be the maximum permissible strain. This is particularly important at elevated temperatures.

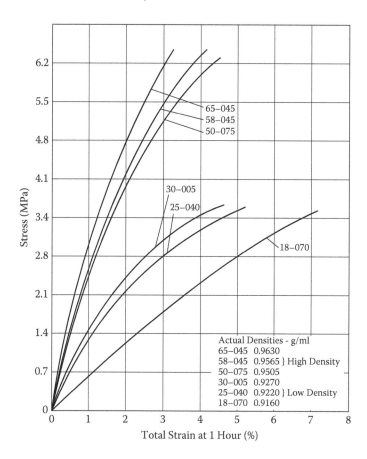

FIGURE 2.8 Isochronous stress–strain curves of various grades of LDPE and HDPE at 23°C (1,000 h data).

The effect of temperature on the creep of polypropylene is also conveniently summarized in Figure 2.11. Although these curves refer specifically to 0.5% and 2% total strain in 1,000 h, similar interpolations have shown that to a first approximation, the fall in stress with temperatures is similar to that for other strain–time combinations.

In contrast to polyethylene, an excellent correlation has been found between the creep and melt index of polypropylene. An increase in the melt index is associated with an increase in creep resistance.

2.5.5 Long-Term Strain–Time Data

To illustrate the linearity of double-logarithmic plots of percentage total strain against time, Figure 2.12 shows a typical family of curves at a range of stress levels for a high-density polyethylene. From the uniform pattern of behavior, the linear extrapolation of the curves at lower stresses to longer times appears well justified. As emphasized above, with the higher-melt-index materials at higher temperatures, overextrapolation is dangerous because of the possible onset of rupture at comparatively low strains.

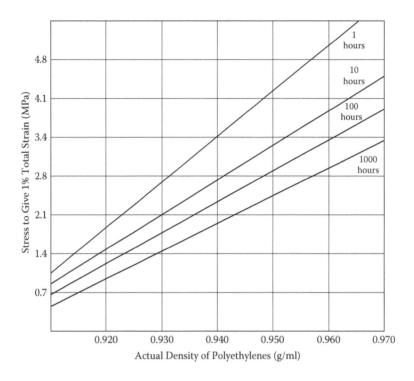

FIGURE 2.9   Effect of density on creep of PE at 23°C.

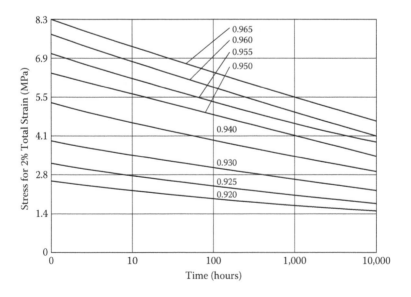

FIGURE 2.10   Interpolated stress–time curves of PE of various densities (2% total strain).

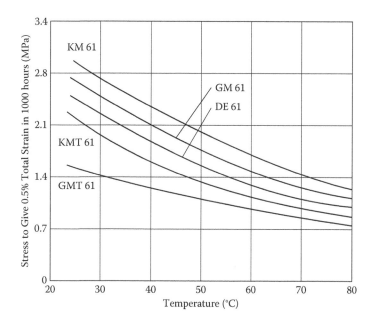

FIGURE 2.11   Interpolated stress–temperature curves of various grades of PP (0.5% total strain at 1,000 h). DE 61 = PP copolymers.

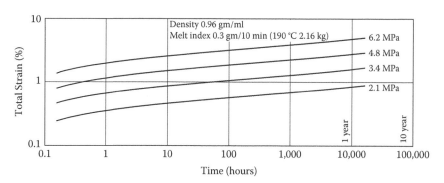

FIGURE 2.12   Long-term creep of HDPE at 23°C.

2.6   IMPACT STRENGTH CHARACTERISTICS OF POLYMERS

Of the many properties of a plastic that influence its choice for a particular article or application, the ability to resist the inevitable sharp blows and drops met in day-to-day use is one of the most important. The prime object of impact testing should be to give a reliable guide to practical impact performance. However, the performance requirements and the design and size of article can vary considerably. The method of fabrication can also vary, and because all these factors can influence impact performance, a reasonably wide range of data are required if most of these eventualities are to be covered and materials and grades compared sensibly.

TABLE 2.5

Impact Resistance (Izod Charpy) and Notched Impact Resistance

	ATS FAAR Number	Units	Suitable Test
Izod Charpy Impact Test			
Pendulum impact strength	16.10000	ft.lb/in.	ASTM D26-56 [44]
(Izod and Charpy) and notched	16.10500	kj/m^2	DIN EN ISO [179] [21]
Impact strength	16.10600	Nm/m	ISO 180 [46]
			BS 2782 [32]
			UNI EN ISO [46]
Falling Dart Impact Test (Fall-O-Scope ATS FAAR)			
Impact resistance (free falling) (dart)			ASTM D5420 [47]
			ASTM D5628 [48]
			DIN EN ISO 6603 [49]
Guided Universal Impact Test (Fall-O-Scope ATS FAAR)			
Impact resistance at low temperature (cable)			

Available instrumentation is listed in Table 2.5.

2.6.1 NOTCHED IZOD IMPACT STRENGTH

Of all the types of mechanical test data normally presented for grade characterization and quality control purposes, the standard notched Izod test is probably the most vigorously criticized on the grounds that it may give quite misleading indications regarding impact behavior in service. In most cases, however, the criticism should be leveled at too wide an interpretation of the results, particularly the tendency to overlook the highly important influence of the differences between the standard and operational conditions in terms of dimensions, notched effects, and processing strains. For example, in the Izod test, the specimens are of comparatively thick section (e.g., 6.25 × 1.25 × 1.25 cm or 6.25 × 1.25 × 0.625 cm) and are consequently substantially unoriented. Even with injection-molded specimens (where some orientation does exist), specimens are usually tested in their strongest direction (with the crack propagating at a right angle to the orientation direction), which generally leads to high values. Furthermore, specimens are notched with the specific intention of locating the point of fracture and ensuring that as wide a range of materials as possible fracture in a brittle manner in the test. It is thus reasonably safe to infer that if a material breaks in a ductile manner at the high-impact energies involved in the Izod test, it will, in nearly all circumstances involving thick sections, behave in a ductile manner in practice. Because of the severe nature of the test, it would be wrong, however, to assume that because a material gives a brittle-type fracture in this test, it will necessarily fail in a brittle manner at low-impact energies in service.

In cases in which high residual orientation is present, additional weakening can result from the easier propagation of cracks in the direction parallel to the orientation.

A low Izod value does give a warning that care should be exercised in design to avoid sharp corners or similar points of stress concentration. It does not by itself, however, necessarily indicate high notch sensitivity.

2.6.2 Falling Weight Impact Test

The normal day-to-day abuse experienced by an article is more closely simulated by the falling weight impact–type test (see BS 2782) [32]. It is also much easier to vary the type and thickness of specimens in this test. Furthermore, any directional weakness existing in the plane of the specimens is easily detected.

ATS FAAR supplies the Fall-O-Scope Universal apparatus for free-falling dart impact resistance tests. The apparatus can be set to obtain the rate of energy absorption during the impact. The apparatus can be used to conduct tests to different specifications with computerized operation in the range 70°C–200°C.

One would normally expect the impact energy required for failure to increase with specimen thickness, but the rate of increase will not be the same for all materials. In addition, the effect of the fabricating process cannot be overemphasized. It is mentioned above that residual orientation can lead to marked weakening and susceptibility to cracking parallel to the orientation direction. Such residual orientation is extremely likely to be brought about in the injection molding process, where a hot plastic melt is forced at a high shear rate into a narrow, relatively cold cavity. Naturally, the more viscous the melt and the longer and narrower the flow path, the greater the degree of residual orientation expected. Within a given family of grades, the melt viscosity, besides being influenced by temperature and shear rates, will also be affected by polymer parameters such as molecular weight and molecular weight distribution. Thus, it cannot be assumed that the magnitude of this drop in impact strength will be the same for differing materials or for differing grades of the same material. For a satisfactory comparison, it is therefore essential to know the variation in impact strength with the thickness of unoriented compression-molded samples and injection-molded samples molded over a range of conditions.

In some investigations, a flat, circular, center-grated, 15 cm diameter molded dish has been used. The dish is mounted in the falling weight impact tester in its inverted position and located so that the strike hits the base 1.87 cm from the center because the area near the sprue is one of the recognized weak spots of injection molding. By means of inserts in the mold, the thickness of the base of the molding can be varied. For each material a family of curves is obtained for variation of impact strength, with the thickness of compression-molded sheets molded under standard conditions and of injection-molded dishes produced at a series of temperatures. An estimate of the weakening effects of residual orientation and of the tolerance that can be allowed on injection molding conditions for satisfactory impact performance is thus obtained.

Besides the average level at which failures occur in the falling weight impact test, the type of failure can also be of importance in judging the relative impact performance of a material in practice. There are generally three types of failure:

- *Ductile or tough failure*, in which the material yields and flows at the point of impact, producing a hemispherical depression, which at sufficiently high impact energies eventually tears through the complete thickness
- *Brittle failure*, in which the specimen shatters or cracks through its complete thickness with no visible signs of any yielding having taken place prior to the initiation of the fracture (with specimens possessing a high degree of residual orientation, a single crack parallel to the orientation direction is obtained)
- *Intermediate or bructile failure*, in which some yielding or cold flow of the specimen occurs at the point of impact prior to the initiation and propagation of a brittle-type crack (or cracks) through the complete thickness of the specimen

An idea of the impact energy associated with these types of failure can be ascertained by consideration of a typical load–deformation curve. This in general will be of the form shown in Figure 2.13. From the definitions given above, a brittle-type failure will occur on the initial, essentially linear part of the curve, prior to the yield point.

A bructile failure will occur on that part of the curve near to but following the yield point. A ductile failure will be associated with high elongations on the final part of the curve, which follows the yield point. With the last type of failure, the deformation at failure will be reasonably constant for a strike of given diameter. The area beneath the curve up to failure gives a measure of the impact energy to cause failure. Hence, in general, for a given thickness, tough-type failure is associated with high and reasonably constant impact energy. Bructile failure is also associated with relatively high but more variable energy. Brittle failures are generally associated with a low impact level.

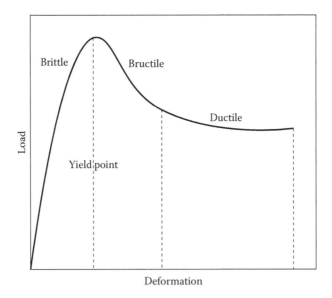

FIGURE 2.13 Typical load-deformation curve.

If, in a given test, failures are all of the tough type, a reliable and reproducible measure of the impact strength of the sample can be obtained. If, however, differing types of failures are observed, a much wider variation in impact values can be obtained from repeat tests.

In practice the lower level at which failure is likely to occur is important as the level at which 50% of the samples fail, i.e., F_{50}, which is the value reported in the falling weight impact test. This can be obtained by repeating the falling weight test using the probit rather than the staircase procedure.

2.6.3 GARDNER IMPACT TEST

Lavach [50] discussed the factors that affect results obtained by the Gardner impact test. This test is used by plastics producers to approximate the mean failure energy for many plastics. The test is inexpensive and easy to operate. The test equipment can be placed close to the manufacturing equipment, permitting fast and nearly online determination of the impact resistance of an article. The test is useful for finding the mean failure energy for brittle thermoplastics such as acrylic and high-impact polystyrene with standard deviations between 8% and 10%.

The sharpness of a notch is well known to have a strong influence on the impact performance of notched specimens. The extent of this effect is found to vary considerably from polymer to polymer.

As mentioned above, a low Izod value gives a warning regarding the care that should be exercised in design to remove points of stress concentration if optimum impact behavior is to be obtained with a component in practice. It was also mentioned that a more reliable measure of notch sensitivity can be obtained by carrying out tests using a series of notch radii rather than a single standard notch.

Figure 2.14 shows the relative notch sensitivities of some polyolefins. Among the polyolefins, the low-melt-index polypropylene copolymer (GMT 61) has the lowest notch sensitivity, whereas the higher-melt-index polyethylene and polypropylene homopolymers have the highest.

Figure 2.15 shows the variation in standard Izod impact strength (1 mm notch tip radius) with temperature for the same range of materials. By far the most temperature sensitive is the low-melt-index polypropylene (PP) copolymer. It is interesting to compare the sharp fall-off in notched impact strength with this material between 23°C and 0°C with the steady values obtained for falling weight impact strength for the same temperature range (Figure 2.16). The removal of points of stress concentration is thus the most vital factor governing the impact performance at temperatures of 90°C and below.

2.7 SHEAR STRENGTH

Apparatus for the measurement of this property according to ASTM D732 [28] is available from ATS FAAR (Table 2.6). Briatico-Vangosa and coworkers [52] discussed the relationship between surface roughness and shear strength in ultra-high-molecular-weight polyethylene (PE).

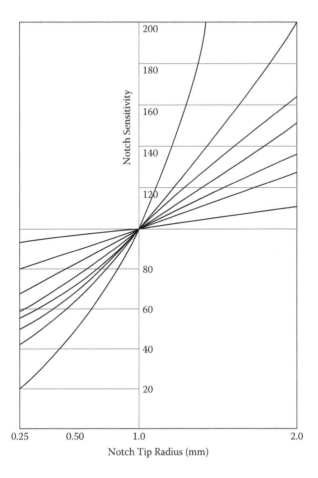

FIGURE 2.14   Typical notch sensitivities of some thermoplastics at 23°C.

2.8   ELONGATION IN TENSION

Apparatus for the measurement of this property according to ASTM D32 [57] are available from ATS FAAR (Table 2.6).

2.9   DEFORMATION UNDER LOAD

The use of light scattering and X-ray beams to measure deformation has been reported by Lee and coworkers [58] and Reikel and coworkers [59].

2.10   COMPRESSIVE SET (PERMANENT DEFORMATION)

The equipment described in the following section is supplied by ATS FAAR (Table 2.6).

Tests can be carried out under constant load Method A according to ASTM D395-03 [53] or under constant deflection according to ASTM D395 [58], ISO 815 [54], or UNI 6121 [55].

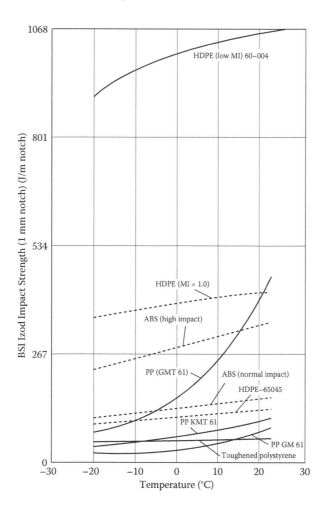

FIGURE 2.15 Influence of temperature on Izod impact strength of various polymers of low and high melt indexes. GMT 61 is a low-melt-index and low-notch-sensitivity polypropylene copolymer. KMT 61 is a high-melt-index and low-notch-sensitivity polypropylene polymer. HDPE 65045 is a high-melt-index and high-notch-sensitivity high-density polyethylene.

The test is used to determine the capability of rubber compounds and elastomers in general to maintain their elastic properties after prolonged action of compression stresses. According to Method A of ASTM D395 [53], this stress is applied by a constant load.

2.11 MOLD SHRINKAGE

Apparatus for the measurement of this property according to DIN 53464 [56] are available from ATS FAAR (Table 2.6). Beracchi and coworkers [60] reported on computer-simulated mold shrinkage studies on talc-filled polypropylene, glass-reinforced polyamide, and polycarbonate/acrylonitrile-butadiene-styrene blends.

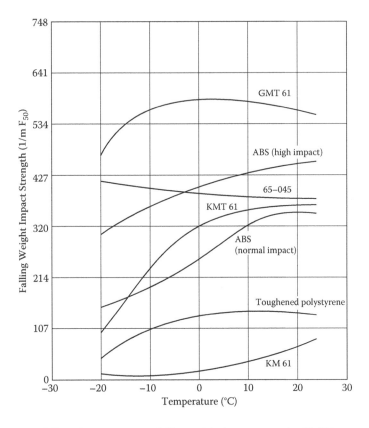

FIGURE 2.16   Effect of temperature on falling weight impact strength of 0.06 in. compression mold specimens of various polymers. PP GMT 61 is a low-melt-index and low-notch-sensitivity polypropylene copolymer. PP KMT 61 is a high-melt-index and low-notch-sensitivity polypropylene copolymer. HDPE 65045 is a high-melt-index and high-notch-sensitivity high-density polyethylene.

TABLE 2.6
Instrumentation, Compressive Set, and Mold Shrinkage

Method	ATS FAAR Apparatus Code Number	Units	Test Suitable for Meeting the Following Standards
Compressive set to measure permanent deformation	10.65000 (stress)	—	ASTM D395 [53]
	10.07000 (strain)		ISO 815 [54]
			UNI 6121 [55]
Mold shrinkage injection compression injection compression and postshrinkage	—	%	DIN 53464 [56]

2.12   COEFFICIENT OF FRICTION

This is an assessment of the coefficient of friction of materials in terms of dynamic (sliding) friction against steel. Friction is influenced by temperature, surface contamination, and most important, the two material surfaces:

- An excellent rating indicates a low coefficient of friction.
- A very poor rating indicates a high coefficient of friction.

Miyata and Yamaoka [152] used scanning probe microscopy to determine the microscale friction force of silicone-treated polymer film surfaces. Polyurethane acrylates cured by an electron beam were used as polymer films. The microscale friction obtained by scanning probe microscopy was compared with macroscale data, such as surface free energy as determined by the Owens–Wendt method and the macroscale friction coefficient determined by the ASTM method. These comparisons showed a good linear relationship between the surface free energy and friction force, which was insensitive to the nature of polymer specimens or to silicone treatment methods. Good linearity was also observed between the macroscale and microscale friction force. It was concluded that scanning probe microscopy could be a powerful tool in this field of polymer science. Evrard et al. [153] reported coefficient of friction measurements for nitrile rubber. Frictional properties of polyacetals, polyesters, polyacrylics [63], reinforced and unreinforced polyamides, and polyethylene terephthalate [52] have also been studied.

2.13   FATIGUE INDEX

This is an assessment of the ability of material to resist oscillating (or dynamic) load or deflection-controlled deformation. ATS FAAR supplies flexing machines (De Mattia type) for the measurement of resistance to dynamic fatigue as measured by ASTM Standard methods D430-96 [64] and D813-95 [65].

These test methods are utilized to test the resistance to dynamic fatigue of rubber-like materials if subjected to repeated bending. The tests simulate the stresses in tension or in compression of inflection or in a combination of the three modes of load application to which the materials will be subjected when in use.

Lin and coworkers [41] also investigated the static tensile strength and fatigue behavior of long glass-reinforced semicrystalline polyamide and amorphous polycarbonate composites. The static tensile measurement at various temperatures and tension–tension fatigue loading tests at various levels of stress amplitudes were studied.

2.14   TOUGHNESS

Toughness tests are traditionally carried out at 20°C (–68°F) or 40°C (68°F). This quantity is related to the notched Izod impact strength. It includes an assessment of the overall fracture toughness of the polymer.

2.15   ABRASION RESISTANCE OR WEAR

This is an assessment of the rate at which material is lost from the surface of the specimen when abraded against a steel face. Wear rates are dependent upon numerous factors, including contact pressure, relative velocity, temperature, and initial surface roughness. This rating considers the inherent ability of a material to resist wear.

The mechanisms of wear are abrasion, adhesion, erosion, fatigue, and fretting. Abrasive wear has a contribution of at least 60% of the total cost due to wear [63]. Abrasive wear is caused by hard particles that are forced and moving along a solid surface [67]. In design, there are two main characteristics that make polymers and reinforced polymers attractive compared to conventional metallic materials. These are relatively low density and reliable tailoring capability to provide the required strength and stiffness. In fact, the abrasion involves the tearing away of small pieces of materials; therefore, the tensile strength, fatigue life, and hardness are important factors in determining the wear characteristics of a polymer. Having said that, there is a need to understand the basic phenomena of two- and three-body absorption [68–72] and the movement pattern of the dry and loose abrasive particles [73]. Much work has been carried out on abrasive wear mechanisms of polymers in general and polymer composites in particular [74–78]. Most test programs have used two-body abrasion tests. In a review of some of the literature concerning the abrasive wear of polymers, Evans and Lancester [79] tested 18 polymers; low-density polyethylene exhibited the lowest wear rate in abrasion against a rough mild steel, but the highest wear rate was in the case of abrasion with coarse corundum paper. Shipway and Ngao [82] investigated the abrasive behavior of polymeric materials on a microscale level. They concluded that the wear behavior and wear rates of polymers depended critically on the polymer type. Furthermore, the wear was associated with indentation-type morphology in the wear scar and low values of tensile strain to failure. Harsha and Tewari [68] investigated the abrasive wear behavior of polyaryletherketone and its composites against SiC abrasive paper. They concluded that the sliding distance load and abrasive grit size have a significant influence on abrasive wear performance. Liu et al. [81] investigated the abrasive wear behavior of ultra-high-molecular-weight polyethylene polymer. They concluded that the applied load is the main parameter, and the wear resistance improvement of filler-reinforced ultra-high-molecular-weight polyethylene was attributed to the combination of hard particles, which prevents the formation of deep, wide, and continuous furrows. Bijwe et al. [82] and Xu and Mellor [83] tested polyamide 6, polytetrafluoroethylene (PTFE), and their various composites in abrasive wear under dry and multipass conditions against silicon carbide paper on a pin-on-disc arrangement. They concluded that the polymers without fillers had better abrasive wear resistance than their composites. Rajesh et al. [84] studied the behavior of polyamide 6,6. They concluded that the water absorption and thermal properties affected the morphology of polyamides, which in turn affected the tribological properties of polyamides. Furthermore, the specific wear rates showed fairly good correlation with various mechanical properties, such as ductility, fracture surface energy, tensile modulus, and the time to failure under tensile stress. Rajesh et al. [85] also investigated the influence of fillers on abrasive wear and polyamide composites. They compared polymer composite wear with wear on unfilled polymers. They evaluated the results by tensile strength and elongation to break properties of the materials. Apart from experimental studies, several models have also been proposed that attempt to relate the abrasive wear resistance of polymers to some mechanical properties of the material, such as hardness and tensile strength.

Ozel [86] studied the abrasive wear behavior of liquid crystal polymer, 30% reinforced polyamide 4,6, and 30% glass fiber–reinforced polyphenylene sulfide engineering polymers at atmospheric conditions. Pin-on-disc arrangement wear tests were carried out at 1 m/s test speed and load values of 4, 6, and 8 N. Test durations were for 50,100 m and 150 m sliding distances.

The specific wear rates of liquid crystalline polymers, 30% glass fiber–reinforced polyamide 4,6, and 30% glass fiber–reinforced polyphenylene sulfide ranged from 3×10^{-2} to 4.43×10^{-1} mm^3/Nm, 1.63×10^{-2} to 1.1×10^{-1} mm^3/Nm, and 2.4×10^{-2} to 2.1×10^{-1} mm^3/Nm, respectively.

The specific wear rates are influenced by the mechanical properties of material, such as tensile strength × elongation to break value. The higher is this value, the lower is the wear rate.

Demir [87] studied the dry sliding wear behavior of polyetherimide/20% glass fiber–reinforced composites and polysulfone/20% glass fiber–reinforced composites used in electrical contact breaker components using a pin-on disc rig. Disc materials employed were steel and polyamide 4,6/30% glass fiber–reinforced composites and wear tests conducted at 0.5 and 1.0 m/s sliding speeds and 20, 40, and 60 N load values under atmospheric conditions of temperature and humidity. Different combinations of rubbing surface were examined and the dynamic friction coefficients and specific wear rates determined and compared. Polymer pin worn surfaces were examined by optical microscopy and wear mechanisms identified as being a combination of adhesive and abrasive wear.

For all inertial combinations, the coefficient of friction showed little sensitivity to sliding speed and applied load value and large sensitivity to material combinations. For specific wear rate, polyetherimide-glass fiber composite showed little sensitivity to the change in load, speed, and material combination, while polysulfone composites showed large sensitivity to the change in load and material combinations. The friction coefficient of 20% glass fiber–reinforced polyetherimide and 20% glass fiber–reinforced polysulfone rubbed off against a steel disc was about 0.3. It was reinforced with 30% glass fiber.

The specific wear values from polyetherimide and polysulfone composites were in the order of 10^{-15} to 10^{-14} mm^3/Nm.

Tabulation of fatigue index, abrasion resistance, and coefficient friction is useful if combined with the major mechanical requirement of the polymer in reaching a decision on compromises that need to be made when choosing a polymer that is suitable for a particular application. Polymers that have a combination of a very good category for all three of these characteristics include high-density polyethylene and polypropylene [63, 64], various polyimide fluoroethylenes, and polyvinylidene fluoride.

The incorporation of 20% glass fiber into polyphenylene oxide leads to an improvement in wear resistance, but has a beneficial effect on either fatigue or friction properties. On the other hand, the incorporation of 20% glass fiber into perfluoroalkoxyethylene leads to an improvement in the coefficient of friction without any beneficial effect or fatigue on wear properties.

2.16   APPLICATION OF DYNAMIC MECHANICAL ANALYSIS

2.16.1   THEORY

Dynamic mechanical analysis provides putative information on the viscoelastic and rheological properties (modulus and damping) of materials. Viscoelasticity is the characteristic behavior of most materials in which a combination of elastic properties (stress proportional to strain rate) is observed. Dynamic mechanical analysis simultaneously measures the elastic properties (modulus) and viscous properties (damping) of materials.

Dynamic mechanical analysis measures changes in mechanical behavior, such as modulus and damping as a function of temperature, time, frequency, stress, or combinations of these parameters. The technique also measures the modulus (stiffness) and damping (energy dissipation) properties of materials as they are deformed under periodic stress. Such measurements provide quantitative and qualitative information about the performance of materials. The technique can be used to evaluate reinforced and unreinforced polymers, elastomers, viscous thermoset liquids, composite coating and adhesives, and materials that exhibit time, frequency, and temperature effects or mechanical properties because of their viscoelastic behavior.

Some of the viscoelastic and rheological properties of polymers that can be measured by dynamic mechanical analysis are [90–92]

- Tensile modulus and strength (elastic properties) [100–103]
- Viscosity (stress–strain rate)
- Prediction of impact resistance [47, 48]
- Damping characteristics
- Low- and high-temperature behavior (stress–strain) [103–107]
- Viscoelastic behavior
- Compliance
- Stress relaxation and stress relaxation modulus
- Storage modulus [91–97]
- Creep
- Deflection temperature under load (DTUL)
- Gelation
- Projection of material behavior
- Polymer lifetime prediction
- Elastomer low-temperature properties

Basically, this technique involves measurement of the mechanical response of a polymer as it is deformed under periodic stress. It is used to characterize the viscoelastic and rheological properties of polymers.

Dynamic mechanical analysis provides a measurement of the mechanical response of a material as it is deformed under periodic stress. Material properties of primary interest include modulus (E'), loss modulus (E''), tan (E''/E'), compliance, viscosity, stress relaxation, and creep. These properties characterize the viscoelastic performance of a material.

Dynamic mechanical analysis provides material scientists and engineers with the information necessary to predict the performance of a material over a wide range of conditions. Test variables include temperature, time, stress, strain, and deformation frequency. Because of the rapid growth in the use of engineering plastics and the need to monitor their performance and consistency, dynamic mechanical analysis has become the fastest-growing thermal analysis technique.

Recent advances in material sciences have been the cause and effect of advances in the technology of materials characterization.

The theory of dynamic chemical analysis has been understood for many years. However, because of the complexity of measurement mechanics and the mathematics required to translate theory into application, dynamic chemical analysis did not become a practical tool until the late 1970s, when DuPont developed a device for reproducibly subjecting a sample to appropriate mechanical and environmental conditions. The addition of computer hardware and software capabilities several years later made dynamic mechanical analysis a viable tool for industrial scientists because it greatly reduced the analysis times and labor intensity of the technique.

Dynamic mechanical analysis provides information on the viscoelastic properties of materials. Viscoelasticity is the characteristic behavior of most materials in which a combination of elastic properties (stress proportional to strain) and viscous properties (stress proportional to strain rate) is observed. Dynamic mechanical analysis simultaneously measures elastic properties (modulus) and viscous properties (damping) of a material. Such data are particularly useful because of the growing trend toward the use of unreinforced or reinforced polymeric materials as replacements for metal and structural applications.

The DuPont 9900 system controls temperature and collects, stores, and analyzes the resulting date. Date analysis software for the DuPont 98/9900 DMA system provides hard-copy reports of all measured and calculated viscoelastic properties, including time–temperature superpositioning of the data for the creation of master curves. A separate calibration program automates the instrument calibration routine and ensures highly accurate reproducible results. In addition to measuring the specific viscoelastic properties of interest, the four modes provide a more complete characterization of materials (including their structural and end-use performance properties). The system can also be used to simulate and optimize processing conditions, such as those with thermoset resins.

The DuPont 983 DMA system can evaluate a wide variety of materials, ranging from the very soft (e.g., elastomers, supported vicious liquids) to the very hard (e.g., reinforced composites, ceramics, metals). The clamping system can accommodate a range of sample geometries, including rectangles, rods, films, and supported liquids. The DuPont 983 instrument (Figure 2.17) produces quantitative information on the viscoelastic and rheological properties of a material by measuring the mechanical response of a sample as it is deformed under periodic stress.

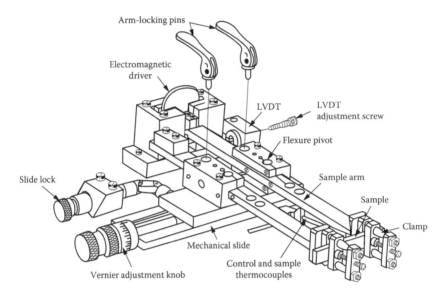

FIGURE 2.17　Electromechanical systems. Variable differential transformer.

Dynamic mechanical analysis can be used in several modes in the evaluation of unreinforced or reinforced plastics.

2.16.2　Fixed-Frequency Mode (i.e., Time–Stress Plots)

This mode is used for accurate determination of the frequency dependence of materials and prediction of end-use product performance. In the fixed-frequency mode, applied stress (i.e., force per unit area that tends to deform the body, usually expressed in Pa (N/m)) forces the sample to undergo sinusoidal oscillation at a frequency and amplitude (strain), that is, deformation from a specified reference state, measured as the ratio of the deformation to the total value of the dimension in which the strain occurs.

2.16.3　Resonant Frequency Mode (i.e., Time–Oscillation Amplitude)

This mode is used for detection of subtle transitions, which are essential for understanding the molecular behavior of materials and structure–property relationships. Allowing a sample to oscillate at its natural resonance provides higher damping sensitivity than at a fixed frequency, making possible the detection of subtle transitions in the material. Because of its high sensitivity, the resonance mode is particularly useful in analyzing polymer blends and filled polymers, such as reinforced plastics and composites.

2.16.4　Stress Relaxation Mode (i.e., Time–Percent Creep Plots)

Stress relaxation is defined as a long-term property measured by deforming a sample at a constant displacement (strain) and monitoring decay over a period of time. Stress is a very significant variable that can dramatically influence the properties of plastics and composites.

2.16.5 CREEP MODE (I.E., TEMPERATURE–FLEXURAL STORAGE MODULUS PLOTS)

Creep is defined as a long-term property measured by deforming a sample at a constant stress and monitoring the flow (strain) over a period of time. Viscoelastic materials flow or deform if subjected to loading (stress). In a creep experiment, a constant stress is applied and the resulting deformation measured as a function of temperature and time. Just as stress relaxation is an important property to structural engineers and polymer scientists, so is creep behavior.

The elastic modulus is a quantitative measure of the stiffness or rigidity of a material. For example, for homogeneous isotropic substances in tension, the strain (ε) is related to the applied stress (o') by the equation $E = o'/\varepsilon$, where E is defined as the elastic modulus. A similar definition of shear modulus (g) applies if the strain is shear.

The thermal curve in Figure 2.18 shows the flexural storage modulus and loss properties of a rigid pultruded oriented fiber-reinforced vinyl ester composite. The flexural modulus and T_g increased dramatically with postcuring, so the test can be used to evaluate the degree of cure, as well as to identify the high-temperature mechanical integrity of the composite.

The creep mode is used for the measurement of flow at constant stress to determine the load-bearing stability of materials, which is a key to prediction of product performance. The creep mode is used to measure sample creep (strain) as a function of time and temperature at a selected stress. Using the isothermal step method, the sample

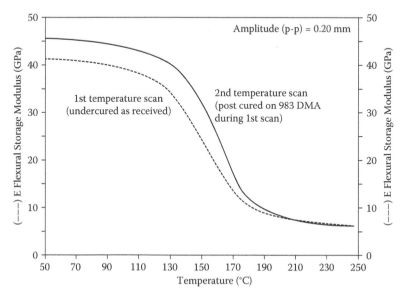

FIGURE 2.18 Flexible storage modulus and loss properties of fiber-reinforced vinyl ester composite.

is allowed to equilibrate at each temperature in a relaxed state. After equilibration, the sample is subjected to a constant stress (Figure 2.18). The resulting sample deformation (strain) is recorded as a function of time for a period selected by the operator. After the first set of measurements is made, the driver stress is removed and the sample is allowed to recover in an unstressed state. Sample recovery (strain) can be recorded as a function of time for any desired period. If the measurement at one temperature is complete, the temperature is changed and the measurement repeated.

Creep studies are particularly valuable for determining the long-term flow properties of a material and its ability to withstand loading and deformation influences.

The task of evaluating new materials and projecting their performance for specific applications is a challenging one for engineers and designers. Often, materials are supplied with short-term test information, such as deflection temperature under load (DTUL) (ASTM D648) [151], which is used to project long-term, high-temperature performance. However, because of factors such as polymer structure, filler loading and type, oxidative stability, part geometry, and molded-in stresses, the actual maximum long-term use temperatures may be as much as 150°C below or above the DTUL. DMA continuously monitors material modulus with temperature, and hence provides a better indication of long-term elevated-temperature performance.

Figure 2.19 illustrates the DMA modulus curves for three resins with nearly identical deflection temperature under load: a polyethylene terephthalate (PET), a polyethersulfone, and an epoxy. The PET begins to lose modulus rapidly at 60°C as the material enters the glass transition. The amorphous component of the polymer achieves an increased degree of freedom, and at the end of the T_g, the modulus of the material has declined by ~50% from room temperature values. Because of its crystalline component, the material then exhibits a region of relative stability. The modulus again drops rapidly as the crystalline structure approaches the melting point. Because the T_g in a semicrystalline thermoplastic is 150°C below its melting

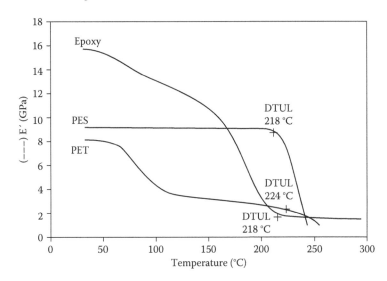

FIGURE 2.19   Comparison of DTUK and DMA results.

point, the actual modulus of a resin of this type at the DTUL is only 10%–30% of the room temperature value. The DTUL of highly filled systems based on these resins is more closely related to the melting point than to the significant structural changes associated with the T_g.

2.16.6 PROJECTION OF MATERIAL BEHAVIOR USING SUPERPOSITIONING

The effects of time and temperature on polymers can be predicted using time–temperature superpositioning principles. The latter are based upon the premise that the process involved in molecular relaxation or rearrangements occurs at greater rates at higher temperatures. The time over which these processes occur can be reduced by conducting the measurement at elevated temperatures and transposing the data to lower temperatures. Thus, viscoelastic changes that occur relatively quickly at higher temperatures can be made to appear as if they occurred at longer times simply by shifting the data with respect to time.

Viscoelastic data can be collected by conducting static measurements under isothermal conditions (e.g., creep or stress relaxation) or by undertaking frequency multiplexing experiments in which a material is analyzed at a series of frequencies. By selecting a reference curve and then shifting the other data with respect to time, a master curve can be generated. A master curve is of great value because it covers times or frequencies outside the range readily accessible by experiment.

Figure 2.20a shows the DMA frequency multiplexing results for an epoxy-fiberglass laminate. The plot shows the flexural storage modulus (E^1) and the loss modulus (8 Hz) as a function of temperature at various analysis frequencies between 0.1 and 8 H_3. The loss modulus peak temperature shows that T_g moves to higher temperatures as the analysis frequency increases. Through suitable calculations, the storage modulus data can be used to generate a master curve (Figure 2.20b). The reference temperature of this is 125°C and shows the effects of frequency on the modulus of the laminate at this temperature. At very low frequencies (or long times), the material exhibits a low modulus and would behave similarly to a rubber. At high frequencies (or short times), the laminate behaves like an elastic solid and has a high modulus. This master curve demonstrates that data collected over only two decades of frequency can be transformed to cover more than nine decades.

Examples of the practical uses for superpositioning are

- Gaskets: To measure flow (creep and stress relaxation) effects, which reduce seal integrity over time
- Force-fit of snap parts: To measure stress relaxation effects, which can lead to joint failure
- Structural beams: To measure modulus drop with time, which leads to increasing beam deflection under load over time
- Bolt plates: To measure the creep of the polymer, which reduces the stress applied by the fastener
- Hoses: To measure the creep of the polymer, which can lead to premature rupture of the hose

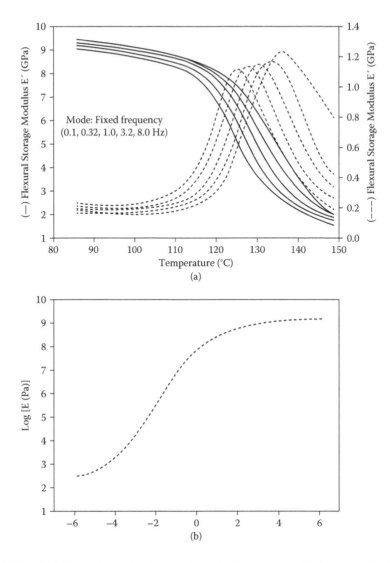

FIGURE 2.20 Dynamic mechanical analysis: (a) frequency multiplexing results for an epoxy-fiberglass laminate and (b) the master curve generated from fixed-frequency multiplexing data for epoxy-fiberglass laminate.

- Acoustic: To aid in the selection of materials that exhibit high damping properties in specific frequency ranges
- Elastomeric mounts: To assess the long-term creep resistance of mounts used for vibration damping with engines, missiles, and other heavy equipment
- Structural parts: To measure high deformation frequencies, which cause shifting of molecular transitions to high temperatures and can result in impact failure or microcracking

Storage modulus data have been reported for polypropylene-wood composites [91], basalt fiber-reinforced polypropylene [92], ethylene-propylene-diene terpolymer [95], polyvinyl fluoride-clay nanocomposites [95], thermosetting resins [95], and water-based adhesives [96, 97].

2.17　RHEOLOGY AND VISCOELASTICITY

As discussed in Section 2.16, dynamic mechanical analysis offers an enhanced means of evaluating the performance of polymeric systems at elevated temperatures. It provides a complete profile of modulus versus temperatures, as well as measurement of mechanical damping. Operating in the creep mode and coupled with the careful use of time–temperature superpositioning, projections can be made regarding the long-term time-dependent behavior under constant load. This provides a much more realistic evaluation of the short- and long-term capabilities of a resin system.

Rheology is concerned with the flow and deformation of matter. Viscoelastic properties are more concerned with the flow and elasticity of matter.

Numerous references to rheology and viscoelasticity occur. Thus, 250 references have been identified in the 5-year period between 2004 and 2008.

In addition to various moduli (storage, loss, loss shear, plateau), the following rheological properties of polymers can be determined by a range of techniques:

- Creep and stress relaxation
- Shear stress
- Stress relaxation behavior
- Order–disorder transitions
- Shear rate dynamic viscosity
- Molecular entanglement
- Creep strain
- T_g
- Melt fracture
- Frequency dependence on dynamic moduli
- Clamping function of shear

Rheological studies have been conducted on several polymers. These include dynamic rheological measurement and capillary rheometry of rubbers [92], capillary rheometry of polypropylene [106], degradation studies on polypropylene [107], torsion rheometry of polyethylene [108], viscosity effects in blends of polycarbonate with styrene-acrylonitrile and acrylonitrile-butadiene-styrene [109], peel adhesion of rubber-based adhesives [110], and the effect of composition of melamine-formaldehyde resins on rheological properties [112].

Particular applications of the techniques in studies of the viscoelastic and rheological properties of unreinforced polymers are reviewed in Table 2.7.

Information on the study of viscoelastic and rheometric properties of reinforced plastics is summarized in Table 2.8.

TABLE 2.7

Measurement of Rheological and Viscoelastic Properties of Unreinforced Polymers

Polymer	Property Measured	Technique	Reference
LLDPE	Storage modulus (alpha relaxation)	Effects of beta irradiation	112
PS	Storage shear modulus	DMTA	113
PBT and polyethylene naphthalate	Stress relaxation behavior	Instron tester	123
Polyvinyl alcohol hydrogels and ferrogels	Storage modulus		114
PEEK (fiber reinforced)	Specimens loaded in flexure	DMA	115
Polystyrene-isoprene	Order transition	Dynamic mechanical spectroscopy	116
PMMA and LDPE	Shear modulus and clamping	Broadband viscoelastic spectroscopy	117
Polybutadienes	Zero shear rate, dynamic viscosity T_g	GPC, FTIR rheometry	118
Polyphenylene sulfide-Fe_3O_4 magnetic composite	T_g, steady state shear deformation	DSC, DMA, TGA	119
Ethylene-propylene random copolymer	Plateau modulus and entanglement	GPC, DSC, oscillatory rheometry	120
PP nanopolymer	Structure and viscoelastic properties	X-ray diffraction, DMA, TGA	121
LLDPE	Creep strain stress relaxation	DSC	122
Polyethylene oxide-polypropylene oxide	Steady-state shear and oscillatory shear	—	123
Branched polyesters	Entanglements	Oscillatory shear measurement	124, 125
Tetrafluoroethylene-hexafluoropropylene copolymer	Zero shear rate viscosity	Linear rheological properties	126
Poly(N-vinylpyrrolidone-co-dimethylamino-propylmethyl-acrylamide)	Radii of gyration	GPC, multiangle light scattering	127
Polyvinyl acetate	Apparent viscosity, viscous flow activation energy	Viscometry	128
Lightly crosslinked acrylic networks	Bulk rheological properties, elastic modulus, resistance to interfacial crack propagation	Dynamic mechanical spectroscopy	129
Polyethylacrylates	Zero rate shear, viscosity dependence	Rheometric measurements	130
Polyalkylether and polyaryletherketone	Dynamic rheological behavior in oscillatory shear mode	Viscosity composition studies	131

TABLE 2.7 (*Continued*)
Measurement of Rheological and Viscoelastic Properties of Unreinforced Polymers

Polymer	Property Measured	Technique	Reference
PVDF	Shear rate	Apparent viscosity activation energy	132
PP modified by ultrasonic waves	Structure of modified PP	High-intensity ultrasonic wave studies	133
Chlorinated polyethylene–natural rubber blends	Viscoelastic changes and thermal degradation	—	134

DMA = dynamic mechanical analysis, DMTA = dynamic mechanical thermal analysis, DSC = differential scanning calorimetry, FTIR = Fourier transform infrared spectroscopy, GPC = gel permeation chromatography, LLDPE = linear low-density polyethylene, PMMA = polymethyl methacrylate, TGA = thermogravimetric analysis.

TABLE 2.8

Rheological and Viscoelastic Properties of Reinforced Plastics

Polymer	Reinforcing Agent	Property Measured	Technique	Comment	Reference
Polycaprolactone	Carbon nanotubes	Nonisothermal crystallization thermal stability	Parallel plate rheometer	To measure rheological percolation of composites under small-amplitude oscillatory shear	134
Polycaprolactone	Multiwalled carbon nanotubes	Nonisothermal crystallization and thermal stability	Parallel plate rheometry, DSC, TGA	Rheological percolation of composites under small-amplitude oscillatory shear	135
Polyethylene ovide	Multiwalled carbon nanotubes	Effect of surface treatment on carbon nanotube dispersion state	Transmission and field emission scanning electron microscopy	Indication of percolated wet work structure	136
Ethylene–vinyl acetate	Multiwalled carbon nanotubes and organically modified clays	Comparison of properties of unmodified and carbon and clay modified polymers	X-ray diffraction	—	137
Nylon 10	Multiwalled carbon	Study of dispersion of carbon nanotubes	Scanning electron microscopy	Carbon nanotubes ast-asa reagent and an increased crystallization	138
Low-density polyethylene	Carbon fiber	Study of rheological properties and oscillatory shear flow	—	Superposed of flow conditions, enhanced dynamic viscosity, and storage modulus	139
Polypropylene	Barium sulfate	Study of effect of surface treatment of barium sulfate on rheological properties	—	Tendency of the polymer chains to undergo intermolecule slippage is received and melt viscosity enhanced by use of coupling agents	140

Low-density polyethylene	Organically modified	Study of effect of fillers on polymer properties	X-ray diffraction, transmission electron microscopy, TGA, DSC, FTIR	—	134
Polyimide-imide	Organically montmorillonite	Study of effect of filler content and type of organic modifier on rheological properties	Intrinsic viscometry	—	141, 142
Polyethylene terephthalate	Silica	Study of effect of silica on rheological and ostallization behavior	—	AV Rami plots for isothermal attained	143
Polymethyl metharcylate	Silica	Study of viscoelastic properties	—	Silica degradation temperature at 10% weight loss was about 3°C higher in presence of silica compared to unfilled polymer influence of filler particles, location and surface treatment on deformation and crystallinity	144
Polyether ether ketone	Calcium carbonate	Study of thermal and mechanical properties	Tensile impact and flexural testing, TGA, DSC, scanning electron microscopy	Influence of filler particles, location and surface treatment on deformation and crystallinity	145
Polyether ether ketone	Fiber glass	Study of effect on glass fiber on rheological properties, including loading on flexure tests	DMA	—	146, 147
Polypropylene	Wood	Measurement of polymer damping peaks. storage modulus, and loss modulus	—	—	93

Continued

31. ISO 458, Plastics—Determination of stiffness on torsion of flexible materials, Part 1: General method, 1985.
32. BS 2782-0, Method of testing plastic parts—Introduction, 1995.
33. ASTM D1869-99, Standard practice for transfer molding test specimens of thermosetting compounds, 2004.
34. ASTM D3419, Standard practice for in-line screw-injection molding test specimens from thermosetting compounds, 2000.
35. ASTM D3641, Standard practice for injection molding test specimens of thermoplastic molding and extrusion materials, 2002.
36. ASTM D4703, Standard practice for compression molding thermoplastic materials into test specimens, plaque, or sheets, 2003.
37. ASTM D5227-01, Standard test method for measurement of hexane extractable content of polyolefins, 2000.
38. H. Wong, P.A. Thompson, J.R. Schoonover, S.R. Ambuchon, and R.A. Palmer, *Polymer Macromolecules*, 2001, 34, 7083.
39. ISO 498, Natural rubber latex concentrated pregnation of dry films, 1992.
40. J. Janick and W. Kroliskowski, *Polimery*, 2002, 47, 250.
41. S.H. Lin, C.C.M. Ma, N.H. Tai, and I. Perng, *Proceedings of the Materials Challenge, Diversification and the Future Symposium*, Anaheim, CA, 1995, 40, 1046.
42. J. Janick and W. Kroliskowski, *Polimery*, 2002, 47, 250.
43. G. Hay, P.E. Mackey, P.N. Awati, and Y. Park, *Journal of Rheology*, 1999, 43, 1099.
44. ASTM D256, Standard test methods for the determination of Izod pendulum impact resistance of plastics, 2005.
45. ISO 180, Plastics, determination of Izod impact strength, 2000.
46. UNIEN ISO 180, Plastics, determination of Izod impact strength, 2001.
47. ASTM D5420, Standard test method for impact resistance for flat rigid plastic specimens by means of striker impacted by a falling weight (Gardner impact), 2004.
48. ASTM D5628, 96 e-l, Standard method for impact resistance of flat rigid plastic specimens by means of a falling part (tup or falling mass), 2001.
49. DIN EN 150 6603, Plastics, determination of puncture impact behaviour of rigid plastics, 2000.
50. M. Lavach, *Plastics Engineering*, 1999, 55, 41.
51. R. Rouabeh, T.R. Fuis, A. Bourdenne, C. Picard, D. Dodache, and H. Houddonic, *Polymer Science*, 2008, 109, 1505.
52. F. Briatico-Vangosa, D. Rinles, F. D'Oria, and A. Verzelli, *Polymer Engineering and Science*, 2004, 40, 15543.
53. ASTM D395-03, Standard method for stiffness properties of a plastic as a function of temperature by means of torsion test, 2003.
54. ISO 815, Rubber, vulcanized or thermoplastic—determination of compression set at ambient elevating or low temperature, 2001.
55. UNI 6121, Elastomer, Provotti Finiti, Prodotti di Gommr Spugnosa, aa Lattice Definizioni E Prove, 1967.
56. DIN 53464, Testing of plastics, determination of shrinkage properties of moulding materials from thermosetting moulding materials, 1962.
57. E. Rouaban, M. Thos, L. Ibos, A. Bourdenne, D. Dadactn, H. Haddaeunis, and I. Ausset, *Journal of Applied Polymer Science*, 2007, 106, 2710.
58. E.C. Lee, M.J. Solomon, and S.J. Muller, *Macromolecules*, 1997, 30, 7313.
59. C. Reikel, M. Burghammer, M.C. Garcia, A. Gourier, and S. Roth, *Polymer Preprints*, 2002, 43, 215.
60. G. Beracchi, A. Ripino, and G. Boero, *Macplas*, 2001, 26, 78.
61. F.L. Jin and S.J. Park, *Journal of Polymer Science, Part B: Polymer Physics*, 2006, 44, 3348.

62. F. Parkes, R. Balart, J. Lopez, and D. Garcia, 2008, *Mech. Compos. Mater.*, 43, 3203.
63. M.Y. Keating, L.B. Malone, and W.D. Saunders, *Journal of Thermal Analysis and Calorimetry*, 2002, 69, 37.
64. ASTM D430-95, Standard test methods for rubber determination, dynamic fatigue, 2000.
65. ASTM D813-95, Standard test method for stiffness properties of plastics as function of temperature by means of torsion test, 2000.
66. M.J. Neale and M. Gee, *Guide Wear Problems and Testing for Industry*, Williams Andrew Publishing, New York, 2001.
67. *Standard Terminology Relating to Wear and Erosion: Annual Book of Standards*, ASTM 1987, p, 243.
68. A.P. Harsha and U.S. Tewari, *Polymer Testing*, 2003, 22, 403.
69. L. Fang, Q.D. Zhou, and Y.J. Li, *Wear*, 1991, 151, 313.
70. K. Grigoroudis and D.J. Stephenson, *Wear*, 1997, 213, 103.
71. B.K. Prasad, S.V. Prasad, and A.A. Das, *Materials Science and Engineering A*, 1992, 156, 205.
72. R.I. Trezona and I.M. Hutchings, *Wear*, 1999, 233, 209.
73. L. Fang, X. King, and Q. Zhou, *Wear*, 1992, 159, 115.
74. A.E. Bolvari and S.B. Gleen, *Engineering Plastics*, 1996, 9, 205.
75. M.A. Moore, in D.A. Rigney (ed.), *Fundamentals of Friction and Wear of Materials*, ASM Metal, 1996.
76. N.P. Sun, Sin, and N. Saka, *Fundamentals of Tribology*, MIT Press, Cambridge, MA, 1980, p. 493.
77. K. Frederick and R. Reinicke, Friction and wear, polymer-based composites, *Mechanical Composite Materials*, 1998, 34, 510.
78. J. Bijwe, U.S. Tewari, and P. Vasudevan, *Wear*, 1989, 132, 247.
79. D.C. Evans and J.K. Lancester, in D. Scott (ed.), *Treatise on Materials Science and Technology*, vol. 13, Academic Press, New York, 1979, p. 85.
80. P.H. Shipway and N.K. Ngao, *Wear*, 2003, 255, 742.
81. C. Liu, L. Ren, R.D. Arnell, and J. Tong, *Wear*, 1999, 225, 199.
82. J. Bijwe, C.M. Logani, and U.S. Tewari, *Proceedings of the International Conference on Wear of Materials*, Denver, CO, April 8–14, 1989, p. 75.
83. Y.M. Xu and B.G. Mellor, *Wear*, 2001, 251, 1522.
84. J.J. Rajesh, J. Bijwe, and U.S. Tewari, *Wear*, 2002, 252, 769.
85. J.J. Rajesh, J. Bijwe, and U.S. Tewari, *Journal of Material Science*, 2001, 36, 351.
86. A Ozel, *Science and Engineering of Reinforced Materials*, 2006, 13, 235.
87. J. Demir, *Journal of Polymer Engineering*, 2009, 29, 549.
88. C. Treney and B. Dupenrang, *Proceedings of the 55th SPE Conference*, Toronto, Canada, 1997, 2805.
89. P.C. Haschke, *Machine Design*, 2001, 73, 61.
90. S.C. Shik, *Popular Plastics and Packaging*, 2003, 48, 55.
91. J. Son, D.J. Gardner, S. O'Neill, and C. Metaxas, *Journal of Applied Polymer Science*, 2003, 89, 1638.
92. M. Botov, H. Betchav, D. Bikiaris, and D. Pananyioton, *Journal of Applied Polymer Science*, 1999, 74, 523.
93. D. Singh, U.P. Malhotra, and J.L. Vats, *Journal of Applied Polymer Science*, 1999, 71, 1959.
94. L. Priya and J.P. Jog, *Journal of Polymer Science, Part B: Polymer Physics*, 2003, 41, 31.
95. M. Sepe, *Injection Moulding*, 1999, 7, 52.
96. D.W. Bamborough, *Proceedings of the Technomatic Pressure Sensitive Adhesive Technology Conference*, Milan, Italy, 2001, paper 10.
97. D.W. Bamborough, *Proceedings of the Technomatic Pressure Sensitive Adhesive Technology Conference*, Milan, Italy, 1993, section 111, p. 2.

98. D. Dean, M. Husband, and M. Trimmer, *Journal of Polymer Science, Part B: Polymer Physics*, 1998, 36, 2971.

99. E. Kontou and G. Spathis, *Journal of Applied Polymer Science*, 2003, 88, 1942.

100. L. Heux, J.J. Halary, F. Laupretre, and I. Monnerie, *Polymer*, 1997, 38, 767.

101. M.E. Leyva, B.G. Soares, and D. Khastgir, *Polymer*, 2002, 43, 7505.

102. O. Schroder and E. Schmachtenberg, *Proceedings of the ANTEC 2002 Conference*, Dallas, TX, 2001, paper 405.

103. R.A. Palmer, P. Chen, and M.H. Gilson, *Proceedings of the ACS Polymeric Materials Science and Engineering Conference*, Orlando, FL, 1996, 75, p. 43.

104. Y.K. Kim and S.R. White, *Polymer Engineering and Science*, 1996, 36, 2852.

105. G. Sarrkel and A. Choudhury, *Journal of Applied Polymer Science*, 2008, 108, 3442.

106. M. Fujiyama, Y. Kitajima, and H. Inata, *Journal of Applied Polymer Science*, 2002, 84, 2128.

107. B. Vergnes and F. Berzin, *Macromolecular Symposia*, 2000, 158, 77.

108. H. Kwag, D. Rana, K. Cho, J. Rhee, T. Woo, B.H. Lee, and S. Choe, *Polymer Engineering and Science*, 2000, 40, 1672.

109. J. Routon and S.A. Rheo, *Platiques et Elastomers Ragazine*, 2000, 52, 34.

110. A.B. Kummer, *Proceedings of the Pressure Sensitive Tape Council 23rd Annual Technocal Seminar: Pressure Sensitive Tapes for the New Millennium*, New Orleans, LA, 2000, p. 25.

111. M. Doyle and J.A.E. Manson, *Proceedings of the ANTEC 2000 Conference*, Orlando, FL, 2000, paper 169.

112. A.L.A. Bobovich, Y. Unigovski, E.M. Gutman, and E. Kolmakov, *Proceedings of the ANTEC 63rd Annual Conference*, SPE, Boston, 2005.

113. M. Patel, *Polymer Testing*, 2004, 231, 107.

114. P.D.B. Jung, D. Bhattacharyya, and A.J. Easteel, *Journal of Material Science*, 2005, 40, 4775.

115. R. Hernandez, A. Safafian, D. Dopez, and C. Myangos, *Polymer*, 2004, 45, 5543.

116. J.D.O. Mela and R.W. Radford, *Journal of Composite Materials*, 2004, 38, 1815.

117. W. Lifeng, E.W. Cochran, T.P. Lodge, and F.S. Bates, *Macromolecules*, 2004, 37, 3360.

118. J. Capodagli and R. Lakes, *Rheologica Acta*, 2008, 47, 777.

119. O. Robles-Vadquez, A. Gonzales-Alvarez, J.E. Puig, and O. Manero, *Rubber Chemistry and Technology*, 2006, 79, 859.

120. D. Wu, L. Wu, F. Goa, M. Zhong, and C. Yan, *Polymer Engineering and Science*, 2008, 48, 966.

121. J. Ding, X. Ding, X. Riwei. and Y. Ding Sheng, *Journal of Macromolecular Science*, 2005, 1344, 308.

122. K.K. Kar, P. Paik, and J.U. Otaigbe, *Proceedings of the 62nd SPE Annual Conference (ANTEC 2004)*, Chicago, 2004, p. 2495.

123. A.L. Bobovitch, Y. Linigovski, E.M. Gutman, E. Kolmakov, and S. Yyazavkin, *Journal of Applied Polymer Science*, 2007, 103, 3218.

124. J.P. Habas, E. Pavie, A. Happ, and J. Perelasse, *Rheologica Acta*, 2008, 47, 765.

125. J. Hepperle, H. Muenstedt, P.K. Haung, and C.D. Eisenbach, *Rheologica Acta*, 2005, 45, 151.

126. J. Hepperle and M. Muenstedt, *Rheologica Acta*, 2006, 45, 717.

127. J. Strange, S. Waechtex, H. Muenstedt, and H. Kaspar, *Macromolecules*, 2007, 40, 2409.

128. J. Jachowicz, R. McMuller, C. Wu, L. Senak, and D. Koelmel, *Journal of Applied Polymer Science*, 2007, 105, 190.

129. F. Yang, H.-H. Xiong, Q. Wang, and L. Li, *Polymer Materials Science and Engineering*, 2007, 23, 190.

130. A Linder, B. Lestriez, S. Mariot, C. Creton, T. Maevis, B. Luehmann, and R. Brumer, *Journal of Adhesion*, 2006, 82, 267.

131. L. Andreozzi, V. Castelvetro, M. Faetti, M. Giordano, and F. Zulli, *Macromolecules*, 2006, 39, 1880.
132. J. Chan, X. Liu, and D. Yang, *Journal of Applied Polymer Science*, 2006, 102, 4040.
133. J.-L. Zhang, L. Shen, W. Chun Li, and Q. Zheng, *Polymer of Materials Science and Engineering*, 2006, 22, 123.
134. K. Wang, J. Wu, and H. Zeng, *Polymers and Polymer Composites*, 2006, 14, 473.
135. D. Wu, L. Wu, Y. Sun, and H. Zhang, *Journal of Polymer Science, Part B: Polymer Physics*, 2007, 45, 3137.
136. Y.S. Song, *Polymer Engineering and Science*, 2006, 46, 1350.
137. S. Peertrbroeck, M. Alexandre, J.B. Nagy, N. Morean, A. Desfree, F. Monteverde, A. Pulmonb, R. Jerome, and P. Dubois, *Macromolecular Symposia*, 2005, 221, 115.
138. B. Wong, S. Gunn, X. He, and K. Liu, *Polymer Engineering and Science*, 2007, 47, 1610.
139. B. Hausnerova, N. Zelrazilova, T. Kitano, and P. Saha, *Polimery*, 2006, 51, 33.
140. Private communication.
141. Y.C. Kim, S.J. Lee, J.C. Kim, and H. Cho, *Polymer Journal* (Japan), 2005, 37, 206.
142. T. Mikolajczyk and M. Olejnik, *Fine Textiles in Eastern Europe*, 2007, 2, 24.
143. D.W. Chae and R.C. Kim, *Journal of Material Sciences*, 2007, 42, 1238.
144. Y.H. Hu, C.Y. Chen, and C.C. Wang, *Macromolecules*, 2001, 37, 2411.
145. B. Zhou, X. Ji, Y. Sheng, L. Wang, and Z. Jiang, *European Polymer Journal*, 2004, 40, 2357.
146. R. Khoshravan and C. Bathias, Delamination test and EDEB and ENF composite specimens model and mode 2, annual report, VAMAS WGS, 1990.
147. N.V. Abdallah, D. Dean, and S. Campbell, *Polymer*, 2002, 43, 5887.
148. M.E. Leyva, R.G. Svares, and D. Khastgir, *Polymer*, 2002, 43, 7505.
149. A. Valea, M.L. Gonzalez, and A. Mondragon, *Journal of Applied Polymer Science*, 1999, 71, 21.
150. G. Gabriel, E. Kokko, B. Hoefgren, J. Sappaler, and H. Muenstedt, *Polymer*, 2002, 43, 6383.
151. ASTM D648, Standard test method for deflection temperature of plastics under flexural load in the edgewise position, 2007.
152. T. Miyata and Y. Yamaoka, *Kobunshi Ronbunshu*, 2002, 59, 415.
153. G. Evrard, A. Belgrine, F. Corpier, E. Valot, and P. Dang, *Revue Generale de Caoutchouces et Plastiques*, 1999, 777, 95.

3 Mechanical Properties of Reinforced Plastics

There is no doubt that for many critical applications of polymers, reinforcing agents will produce dramatic improvements of mechanical and other physical properties [1–3], such as tensile strength and flexural modulus. In this chapter, work on the effects on physical properties of several types of reinforcing agents is reviewed. We consider the effects of various reinforcing agents on the mechanical properties of a wide range of polymers.

Substances that have been used in this context include glass fiber (occasionally glass beads), carbon fiber, carbon nanotubes, carbon black, graphite, fullerenes, graphite chemically modified clays and montmorillonites, silica, and mineral alumina. Other additions have been included in polymer formulations, including calcium carbonate, barium sulfate, and various miscellaneous agents, such as aluminum metal, oak husks, cocoa shells, basalt fiber, silicone, rubbery elastomers, and polyamide powders. The effects of such additions of polymer properties are discussed next.

3.1 GLASS FIBER REINFORCEMENT

The effect of fiberglass reinforcement on the mechanical properties of polymers is reviewed in Table 3.1. Generally speaking, glass fiber is incorporated in the 20%–50% range. This has an appreciated effect on tensile strength and flexural modulus.

3.1.1 TENSILE STRENGTH

It is seen in Table 3.1 that polyamide 6 improves its tensile strength from a value of 40 MPa to 145 MPa, that is, by a factor of about 3.6. The improvement obtained by the incorporation of glass fiber into polybutylene terephthalate was even more dramatic, from 52 to 80–196 MPa. Very useful improvements in tensile strength result from the incorporation of glass fiber into the formulation. Such improvements ranged from 17 to 41 MPa on unreinforced polymer to 180 MPa on glass-reinforced polytetrafluoroethylene.

Such enhancements in tensile properties by the incorporation of glass fiber into the polymers have been important in the introduction of plastics into engineering applications.

3.1.2 FLEXURAL MODULUS

With the exception of epoxies and perfluoroalkoxy, the incorporation of 25%–50% glass fiber reinforcement leads to an improvement in flexural modulus. As shown in Table 3.2, the outstanding polymers are polyethylene terephthalate, where

TABLE 3.1

Heat of Glass Fiber on Tensile Strength of Polymers

Polymer	% Glass Fiber Reinforcement	Unreinforced (MPa)	Reinforced (MPa)	Improvement Factor (MPa Reinforcement/ MPa Unreinforced)
Epoxy resin	—	34–54	68	0.81–2
Polybutylene terephthalate	43	52	80–196	1.5–3.8
Polyether etherketone	50	92	150	1.6
Polyamide-imide	—	185	195	1.10
Polyethylene terephthalate	30	55–54	100	1.2–1.8
Polytetrafluoroethylene	25	17–41	180	4.4–9.4
Polyphenylene oxide	30	50–65	85	1.3–1.7
Perfluoroalkoxyethylene	20	29	33	1.1
Polyimide	40	72	79	1.11
Polycarbonate	50	50–66	50	0.75–1
Fire-retardant geode epoxy resins	—	600	68	0.1
Polyamide 6	—	40	145	3.6

Previewed from H. Eriksson, *Kunststoffe*, 2009, 9, 88. Springer Verlag © 2009.

TABLE 3.2

Effect of Glass Reinforcement on Flexural Modulus of Polymers

Polymer	% Glass Fiber Reinforcement	Unreinforced (GPa)	Reinforced (GPa)
Polyamide 6	—	1	1.6
Epoxy resin	—	3–3.5	1.1
Polybutylene terephthalate	45	2.3	5–17.9
Polyether ether ketone	50	3.7–3.9	10.3
Polyamide-imide	—	4.6	11.1
Polyethylene terephthalate	30	2.3	9.5
Polytetrafluoroethylene	25	0.7	1.3
Polyphenylene oxide	30	2.5	17.2
Perfluoroalkoxyethylene	20	0.7	0.7
Polyimide	40	2.4	13.2
Polycarbonate	50	2.1–2.8	2.1
Fire-retardant fluoroepoxy	—	80	1.1
Polybutylene terephthalate	20	2.1	5

TABLE 3.3

Glass Fiber–Reinforced Polymers That Combine High Tensile Strength and Flexural Modulus

Polymer	% Glass Fiber	Tensile Strength (GPa)	Flexural Modulus (GPa)
Polyamide-imide	—	185	11.1
Polyether ether ketone	56	150	10.3
Polybutylene terephthalate	45	80–196	5–17.9
Polyethylene terephthalate	—	100	9.5

incorporation of 30% glass fiber improves the flexural modulus from about 2.3 to 9.5 GPa; polyether ether ketone (improvement from 0.3.7 to 10.3 GPa); polyphenylene oxide (improvement from 2.5 to 17.2 GPa); polyimide (improvement from 2.4 to 13.2 GPa); polybutylene terephthalate (improvement from 2.3 to 5–18 GPa); and polyimide-imide (improvement from 4.6 to 11.1 GPa).

Reinforced polymers that combine high tensile strength with high flexural modulus include polybutylene terephthalate, polyether ether ketone, polyamide-imide, and polyethylene terephthalate (see Table 3.3).

3.1.3 ELONGATION AT BREAK

As seen in Table 3.4, in general the incorporation of glass fiber into polymer formulations, not unexpectedly, leads to a large reduction in elongation at break.

3.1.4 IMPACT STRENGTH

As shown in Table 3.5, the incorporation of glass fibers causes little or no change in Izod impact strength.

Published information on a range of reinforced polymers is now discussed.

3.1.5 MECHANICAL PROPERTIES OF REINFORCED POLYOFINS

Long glass fiber–reinforced resins have established a strong presence in the injection molding industry [4]. Optimizing the mechanical properties requires control of fiber orientation during molding. Computer-aided engineering simulations have proved beneficial in mold design, but these simulations depend on the quality and relevance of the rheological properties. The flow behavior of long glass fiber/polypropylene grades measured in a variety of rheometric geometries is presented.

Gupta et al. [2] studied the effect on Izod impact strength of incorporating a chemical coupling agent into fiberglass-reinforced polypropylene.

A maximum improvement in this property above a rate of 0.07 kj/m was obtained for virgin polypropylene at a 1% addition rate of the modifier. A decrease in impact

TABLE 3.4
Effect of Glass Fiber Reinforcement in Elongation at Break of Polymers

Polymer	% Glass Reinforcement	Unreinforced %	Reinforced %
Polyamide 6	—	4.0–6.0	2.4–3.15
Epoxy resin	—	1.3	—
Polybutylene terephthalate	45	250	2–3
Polyether ether ketone	50	50	2.2
Polyamide-imide	—	12	5
Polyethylene terephthalate	30	300	2.2
Polytetrafluoroethylene	25	240–600	240
Polyphenylene oxide	30	20–60	1
Perfluoroalkoxyethylene	20	300	4
Polyimide	40	4–8	1.2
Polycarbonate	50	60–200	200
Fire-retardant grade epoxy	—	1.3	—
Polybutylene terephthalate	20	250	2–3

TABLE 3.5
Effect of Glass Fiber Reinforcement on Izod Impact Strength of Polymers

Polymer	% Glass Reinforcement	Unreinforced (kj/m)	Reinforced (kj/m)
Polyamide 6	—	0.25	0.6
Epoxy resin	—	0.5	0.5
Polybutylene terephthalate	45	0.06	0.03
Polyether ether ketone	50	0.08	0.09
Polyamide-imide	—	0.13	0.10
Polyethylene terephthalate	30	0.02	0.06
Polytetrafluoroethylene	25	0.11–0.16	0.25
Polyphenylene oxide	30	0.16	0.058
Perfluoroalkoxyethylene	20	1.06	0.7
Polyimide	40	0.08	0.24
Polycarbonate	50	0.05	0.05
Polybutylene terephthalate	20	0.06	0.03

strength and elongation at break occurred, however, with increasing the addition of glass fiber in the formulation. The resulting polymer blend had significantly improved crack propagation and impact strength properties with a small reduction in tensile strength.

Union Carbide Chemical & Plastics Co. [5] reports that adding 3% of one or two organosilicone compounds (Ucarsil PC-1A and PC-1B) to GE's recycled Azdel polypropylene/glass fiber sheet in regrind form increases the mechanical properties of the recycled sheet, which contains 38% glass fiber reinforcement. Tensile and flexural strength were increased some 35%–40%, and Izod impact strength increased some 30%.

Eriksson [6] has pointed out that when used in conjunction with glass fibers, polypropylene grades of great rigidity can be obtained. The high rigidity and low density of these highly crystalline polypropylene compounds allow them to act as a substitute for engineering polymers, such as polyamide and polybutylene terephthalate. Performance at elevated temperatures opens up scope for using this polymer in automotive applications that were hitherto the preserve of engineering polymers. It also makes for lighter parts.

A highly crystalline, glass fiber–reinforced, speciality polypropylene, marketed under the name Polyfill PP HC, has been developed by Polykemi AB, Ystad, Sweden. The glass fiber content in the polypropylene compound can be varied and enables the wall thickness to be customized and part weights to be commensurately reduced.

The properties of the polypropylene can be so selectively dimensioned that the highly crystalline glass fiber–filled compound can even act as a substitute for engineering plastics, such as polyamide and polybutylene terephthalate. For example, its rigidity is almost on a par with that of a freshly molded, reinforced polyamide 6. The notched Charpy impact strength of polypropylene is higher than that of freshly molded polyamide 6. When polyamide 6, which is hygroscopic, is conditioned, its impact strength increases too. Again, highly crystalline polypropylene offers superior rigidity, which is reduced in polyamides when they absorb moisture. Polyfill PP HC, like normal polypropylene, is moisture resistant, and so this factor can be ignored when it comes to parts design and mechanical properties. Comparison of polyamide 6 and polybutylene terephthalate, each with a 30% glass fiber content, shows polyamide 6 to be about 20% denser than the polypropylene-polybutylene terephthalate reinforced with a similar content, and up to 35% denser than the polypropylene. Polyfill PP HC has the potential to serve as a replacement for polybutylene terephthalate. Polyfill PP HC containing 5%–10% less glass fiber offers the same rigidity as a traditional polypropylene of higher glass fiber content.

The crystalline structure renders Polyfill PP HC suitable for applications involving permanent exposure to high temperature. For example, a polypropylene grade containing 30% glass fiber reinforcement had heat distortion temperatures (A and B values) of 149°C and 159°C. The experimentally determined Vicat A heat deformation temperature of 166°C is thus higher than the melting point of a traditional polypropylene. The flexural elastic modulus at 80°C has a value of 5,800 MPa. This is the value possessed at room temperature by a traditional polypropylene having the same glass fiber content.

To additionally protect the highly crystalline polypropylene homopolymer from thermal degradation at elevated temperatures, the Polyfill PP HC has included in this formulation a heat stabilizer, which offers protection against thermal aging at 150°C. This is the value specified in automotive standards as the maximum temperature for thermal aging test, in which the material must resist exposure for 1,000 h before it shows signs of decomposition. In the trial, samples were conditioned at 150°C for 2,000 h, and the unnotched Charpy impact strength was determined in each case after 500 h. After 1,800 h, a decrease in impact strength was observed—this took the form not of decomposition, but rather of recrystallization. Total decomposition of the polypropylene was apparent only after 4,000 h of aging.

Gupta et al. [3, 7] also studied the mechanical properties of a series of blends of glass fiber reinforced with varying proportions of low-density polyethylene. They found that flexural strengths and tensile and flexural moduli tended to decrease with increasing percentage of low-density polyethylene copolymer, while there was a marked increase in both tensile and flexural strengths and modulus. Where a significant increase was observed, therefore reversing the trend of decrease in tensile properties on incorporation of the elastomer. The effect of glass blends on the mechanical properties of polypropylene has been studied [8]. Workers have studied the effect of glass fibers on the influence of various factors, such as melt temperature [9], weld line strength [10], and annealing on glass-reinforced polymers [51].

3.1.6 MECHANICAL PROPERTIES OF POLYETHER ETHER KETONE

As seen in Table 3.1, the incorporation of 50% glass fiber into polyether ether ketone (PEEK) increased the tensile strength from 92 to 150 MPa and increased the flexural modulus from 3.7 to 10.3 GPa, while decreasing the elongation at break from 50% to 2.2%.

LNP Engineering Plastics [12] introduced a series of formulations based on polyketone-reinforced glass and carbon fiber with polytetrafluoroethylene (PTFE) as lubricant. These polymers have good impact performances for a wide temperature range; they have high chemical resistance, good hydrolytic stability, and superior resilience.

These polymers are targeted at the automotive, business machine, domestic appliances, and electrical markets.

Khoshravan and Bathias [13] used dynamic mechanical analysis to measure rheological and viscoelastic properties of glass fiber–reinforced polyether ether ketone.

3.1.7 MECHANICAL PROPERTIES OF POLYETHYLENE TEREPHTHALATE

As seen earlier, the incorporation of 30% glass fiber into polyethylene terephthalate increases the tensile strength from 55 to 100 MPa (Table 3.1), while increasing the flexural modulus from 2.3 to 5–17.9 GPa and decreasing the elongation of break from 250% to 2.2% (Table 3.3).

Glass-filled polypropylene has been the subject of several studies involving tensile testing and the measurement of notched Izod impact strength. Gupta and coworkers

[3, 7] observed a maximum improvement of the mechanical and thermal properties in polypropylene with a 1% addition of a chemical coupling agent. However, deterioration in impact strength and elongation at break occurred with increasing contents of glass fiber in the polymer.

R. Keating et al. [14] have studied long-term creep straining in glass fiber–reinforced polyamides for periods up to 10 years. They [14] examined three polymers, polyethylene terephthalate, polyamide 6,6, and H T nylon, of glass transition temperatures 72, 58, and 132°C, respectively, to determine the necessity of high-temperature annealing and the degree of its impact on strain.

The improvement after annealing depends on the polymer type and on whether it is fast or slow when crystallizing at the molding process and whether it has hydrogen-bonded sheets. The physical property changes of these materials observed before and after annealing support the explanation of crystal reorganization through crystallization, free volume reduction through densification, and crystal perfection through better chain packing. The prediction of long-term creep strains of up to 10 years using the time–temperature superposition technique has been practiced over a period of 13 years. The accuracy of the prediction is shown with a confidence level of about 90%.

Koh et al. [15] investigated the mechanical and thermal properties of glass-reinforced polyethylene terephthalate and showed that the molding process used was very important in achieving optimal mechanical properties.

3.1.8 MECHANICAL PROPERTIES OF POLYBUTYLENE TEREPHTHALATE

As seen earlier, the incorporation of 45% glass fibers into polybutylene terephthalate increases the tensile strength from 52 MPa to values between 80 and 196 MPa (Table 3.1), giving it a superior tensile strength to polyethylene terephthalate.

At the same time, the flexural modulus (Table 3.2) increases from 2.3 MPa to values between 5 and 17.9 MPa.

Elongation decreases dramatically from 250% for unreinforced polybutylene terephthalate to 2.3% for 45% glass fiber–reinforced polybutylene terephthalate (Table 3.4). Izod impact strength is virtually unaffected by the incorporation of glass fiber into the formulation.

3.1.9 MECHANICAL PROPERTIES OF POLYAMIDES

The incorporation of glass fiber into polyamide 6 increases the tensile strength from 40 to 145 GPa (Table 3.1), accompanied by an increase in flexural modulus (Table 3.2) from 1.0 to 1.6 MPa and hardly any change in elongation at break (Table 3.4).

Li Huang et al. [16] discussed the mechanical properties of polyamides. Lyons [17] obtained creep rupture data and tensile behavior for glass-filled polyamide and polyphthalamide at 2.3°C–150°C.

Karayannidis et al. [18] studied crack propagation and impact strength of glass-reinforced polyamide 6,6 containing up to 12.5% of the functionalized tri-block copolymer styrene (ethylene-co-butylene) grafted with maleic anhydride

(SCBS-g-MA) containing 2.5, 5, 7.5, and 10% wt% copolymers. These polymers were prepared by melt blending in a single-screw extruder. Emphasis was placed on mechanical properties, in comparison with morphology and thermal properties of the aforementioned sample. Although the amount of maleic anhydride grafted by styrene-ethylene-butylene added to the polyamide was not enough to produce tough materials, a significant increase in the resistance to crack propagation and impact strength was observed in all blends. This behavior was proportional to the amount of triblock polymers added for samples having up to 10%, rubber, while additional amounts seem to have no further effect. A small decrease in tensile strength is also observed. From Fourier transform infrared spectroscopy and differential scanning calorimetry analysis, it was shown that the grafting extent of the triblock polymer to polyamide 6,6 is very low, 35%.

3.1.10 Mechanical Properties of Polyamide Imide

The corporation of glass fiber has little effect on the tensile strength of polyamide-imide from 185 to 195 MPa (Table 3.1) and the flexural modulus from 4.6 to 11.1 GPa (Table 3.2), while the elongation of break decreases from 12% to 5% (Table 3.4).

3.1.11 Mechanical Properties of Polyimides

The incorporation of 40% glass fiber into polyimide increased the tensile strength from 72 to 79 MPa (Table 3.1) and the flexural modulus from 2.46 to 13.2 GPa with hardly any effect on impact strength (Table 3.2). The elongation of break falls from 8% to 1.2% (Table 3.4).

3.1.12 Mechanical Properties of Polyphenylene Oxide

The incorporation of 30% of glass fiber into polyphenylene oxide brings about a relatively small increase in tensile strength from between 50 and 65 MPa for the unreinforced polymer to 85 MPa in the reinforced polymer (Table 3.1). The addition of glass fiber is accompanied by an increase in flexural modulus from 2.5 to 17.2 GPa (Table 3.2) and a dramatic decrease in modulus of elasticity from between 20 and 60% down to 1% (Table 3.4). The incorporation of glass fibers into polyphenylene oxide produces a distinct improvement in the wear resistance of the reinforced polymer accompanied by small improvements in fatigue index [19] and the coefficient of friction [20].

3.1.13 Mechanical Properties of Perfluoroalkoxyethylene

The incorporation of 20% glass fiber into perfluoroalkoxyethylene hardly affects tensile strength (29 MPa vs. 33 GPa) (Table 3.1) and has little effect on Izod impact strength (Table 3.5), but causes a dramatic decrease in elongation at break (from 300% to 4%) (Table 3.4).

The incorporation of glass fiber decreases wear resistance but improves the coefficient of friction of this polymer.

3.1.14 MECHANICAL PROPERTIES OF EPOXY RESINS

The incorporation of glass fiber into epoxy resins improves the tensile strength from as low as 34 to 68 MPa (Table 3.1), accompanied by a reduction of flexural modulus from 3 to 1.1 GPa (Table 3.2).

Gojny et al. [21] and Nischitani et al. [22] have discussed the effect of glass fibers on the mechanical and electrical properties of epoxy resins.

These workers studied glass fiber–reinforced epoxy resin containing various amounts of carbon black and carbon nanotubes or double-walled carbon nanotubes. Evaluation of tensile strength, Young's modulus, fracture toughness, and interlaminar shear strength showed that nanoparticle-reinforced epoxies containing carbon black and carbon nanotubes were suitable for resin transfer molding.

3.1.15 MECHANICAL PROPERTIES OF POLYTETRAFLUOROETHYLENE

As seen in Tables 3.1 through 3.4, the incorporation of 25% of glass fiber increases the tensile strength from values between 17 and 41 MPa to 180 MPa, and the flexural modulus from 0.7 to 13.5 GPa, while reducing the elongation at break from values between 400% and 600% to 400%.

3.1.16 MECHANICAL PROPERTIES OF POLYPHENYLENE SULFIDE

Polyphenylene sulfide containing up to 30% glass fiber has recently become available in thermoformable grades. In addition to its remarkable tensile, flexural, and impact properties, this composite can be formed to a depth of 6 in. [23].

3.1.17 MECHANICAL PROPERTIES OF MISCELLANEOUS POLYMERS

Seldon [24] measured weld line strength for injection-molded specimens of acrylonitrile-butadiene-styrene terpolymer, polyphenylene oxide, talc-filled polypropylene, and glass fiber–reinforced polyamide 6 and polyphenylene sulfide. Flexural, impact, and tensile tests were undertaken, and fracture surface examined by scanning electron microscopy. The effects of holding pressure, injection velocity, melt temperature, and mold temperature on weld line strength were studied. For each parameter setting the weld line strength was measured and compared with the bulk strength via weld line factor, defined as the strength of specimens with a weld line divided by the strength of specimens without a weld line. The highest weld line factors were obtained for unfilled materials molded using high melt temperature and holding pressure and low mold temperature.

Sims et al. [25] reviewed some UK contributions to the development of fiber-reinforced plastics such as those used for glass-reinforced plastic pressure vessels and pipes. He also discussed the need for harmonizing and validating the many variants of property testing methods currently in use.

Chen and Lein [26] have described a proprietary process for the manufacture of pultruded hybrid fiber (glass/carbon)-reinforced unsaturated polyester composites. To investigate the effect of reinforcement type and content on the properties, composites with various fiber contents were fabricated. Flexural strength, flexural modulus, notched Izod impact strength, dynamic storage modulus, dynamic loss tangent, and weight loss of pultruded composites were measured. The aim of the investigation was to obtain the best mechanical and thermal properties and the optimum price composites for pultrusion.

Clay-polyvinylidene fluoride composites have significantly improved storage modulus to the base polymer over the temperature range −100°C to 150°C [27]. Gupta and coworkers [2] and Yan and Jiang [28] reported on the mechanical and thermal behavior of glass fiber–filled or glass bead–filled polyvinylidene-bentonite-clay nanocomposites. These were shown to have a significantly improved storage modulus to the base polymers.

Chen et al. [29] have carried out a detailed study of the methods for the prediction of the long-term behavior of rods of glass fiber–reinforced vinyl ester resin used to reinforce concrete building structure. As a measure of durability, tensile strengths are reported for bars of this material before and after exposure to alkaline solutions. In a typical example, the tensile strength of the exposure of glass fiber–reinforced vinyl ester resin changes from 100% to 80% after 400 days exposure and from 100% to 60% after 800 days of exposure to concrete.

3.2 CARBON FIBER REINFORCEMENT

Available information on the effect of carbon fiber on the mechanical properties of various polymers is now reviewed.

3.2.1 MECHANICAL PROPERTIES OF POLYACETAL

As seen in Table 3.6, the addition of 30% carbon black to polyacetal leads to a useful improvement in tensile strength and flexural modulus but, as might be expected, causes a distinct deterioration in elongation properties.

3.2.2 MECHANICAL PROPERTIES OF POLYCARBONATE

As shown in Table 3.7, incorporation of 30% carbon fiber into polycarbonate more than doubles its tensile strength, with a useful increase in flexural modulus and severe loss of percent elongation.

3.2.3 MECHANICAL PROPERTIES OF POLYAMIDES

The incorporation of 10%–30% carbon fiber into polyamides produces distinct improvements in tensile strength and flexural modulus, as shown for some polyamides in Table 3.8.

3.2.4 MECHANICAL PROPERTIES OF POLYETHERSULFONE

The addition of 30% carbon fiber into polyethersulfone increases its tensile strength 84 to 195 MPa, with an increase in flexural modulus from 2.6 to 15.2 GPa and a corresponding decrease in elongation from 6.0% to 1.8%.

TABLE 3.6

Effect on Mechanical Properties of Addition of Carbon Fiber to Acetal Resin

	Unreinforced	30% Carbon Fiber Addition
Tensile strength, MPa	50–73	85
Flexural modulus, GPa	2.6–2.7	17.2–17.2
Elongation, %	65–75	1
Izod impact strength, kj/m	0.06–0.10	0.04

TABLE 3.7

Effect of 3% Carbon Fiber Addition on Mechanical Properties of Polycarbonate

	Unreinforced	30% Carbon Fiber Reinforced
Tensile strength, MPa	50–66	165
Flexural modulus, GPa	2.1–2.8	13
Elongation, %	60–200	2.7
Izod impact strength, kj/m	0.05	—

TABLE 3.8

Mechanical Properties of Virgin and Carbon Fiber–Reinforced Polyamide 6,6

	Unreinforced	30% Carbon Fiber Reinforced
Tensile strength, MPa	40–84	220
Flexural modulus, GPa	1	18
Elongation, %	40–60	—
Izod impact strength, kj/m	0.25	—
	Virgin	**10% Carbon Fiber**
Tensile strength, MPa	48–84	130
Flexural modulus, GPa	1.2	6.5
Elongation, %	60–300	5
Izod impact strength, kj/m	0.68–1.36	—
Strength at yield	4.5	—

3.2.5 MECHANICAL PROPERTIES OF EPOXY RESINS

Carbon fiber has been used as a reinforcing agent in epoxy resins [30, 31].

Sudbury [32] carried out a detailed study of the dynamic stability of jetliner structures constructed in carbon fiber–reinforced epoxy composite. The jetliner in question, the Airbus A300, had run into heavy turbulence, which weakened the structure with consequent buckling. This was due, at least in part, to a weak matrix caused by incomplete curing of the epoxy resin or a deficiency of the curing agent during the manufacturing process.

3.2.6 MECHANICAL PROPERTIES OF LOW-DENSITY POLYETHYLENE

Hausnerova et al. [33] studied the rheological properties of carbon fiber–filled low-density polyethylene. These workers investigated rheological properties of carbon-filled low-density polyethylene melts undergoing parallel superposed steady and oscillatory shear flows. It was also found that the critical angular frequency where the phase angle becomes 90° and storage modulus decreases sharply to zero. This was mainly dependent on the fiber volume fraction.

3.2.7 MECHANICAL PROPERTIES OF POLYVINYLIDENE FLUORIDE

Vidhate et al. [34] have studied the rheology properties of polymeric composites of carbon nanofibers and polyvinylidene fluoride. These workers used rheometry, dynamic mechanical analysis, and differential scanning calorimetry complex modulus (i.e., ratio of couplex stress and composite strain amplitude, i.e., G^*). Tensile stress values are found to be increased by the addition of carbon nanofibers, but the degree of crystallization was affected. The melting point of the polymer was not affected. In addition, a decrease in the glass transition temperature was observed. Lai et al. [36] studied the mechanical properties of carbon fiber–reinforced polyvinylidene-fluoride tetrafluoroethylene as a function of the degree of cross-linking. Increase in cross-linking produced an improvement in mechanical properties.

3.2.8 MECHANICAL PROPERTIES OF PHENOLIC RESINS

Patton et al. [35] evaluated ablation, mechanical, and thermal properties of vapor-grown fiber-phenolic resin (Pyrograf III from Applied Sciences, Inc. and SC-1008 from Borden Chemical, Inc.) composites. The workers determined the potential of using this material in solid rocket motor nozzles. Composite specimens with varying vapor-grown fiber loading (30–50 wt%), including one sample with ex-rayon carbon fiber piles, were prepared and exposed to a plasma torch for 20 s with a heat flux of 16.5 MW/m^2 at about 1,650°C. Low erosion rates and little char formation were observed, confirming that these materials were promising for use as rocket motor nozzle materials. When fiber loadings increased, mechanical properties and ablative properties improved. The vapor-grown carbon fiber composites had low thermal conductivities (about 0.56 W/m-K), indicating that they were good insulating materials. If a 65% fiber loading in vapor-grown carbon fiber composites could be achieved, then ablative properties were expected to be comparable with or better than the composite material currently used on the space shuttle reusable solid rocket motor.

3.2.9 MECHANICAL PROPERTIES OF SYNTACTIC FOAM

Wouterson et al. [37] examined the effect of the carbon fiber content and fiber length on tensile, fracture, and thermal properties of syntactic foam. They showed that the hybrid structure demonstrates a significant increase in the ultimate tensile strength, sigma "u" t" s, and Young's modulus, E, with increasing fiber loading. Interestingly, the fracture toughness, K "I" c, and energy release rate, G "I" c, increased by 95% and 90%, respectively, upon introduction of 3 wt% short carbon fibers in syntactic foam, indicating the potent toughness potential of carbon fibers in syntactic foam systems. Scanning carbon microscopy studies identified the presence of several toughness mechanisms. An estimate of the contribution from each toughening mechanism by composite theory and fractography revealed that the specific energy required to create new surfaces was enhanced by the presence of fibers and was the main contributor to the toughness of the short fiber–reinforced syntactic foam.

3.3 CARBON NANOTUBE REINFORCEMENT

The discovery of carbon nanostructured materials has inspired a range of potential applications. More specifically, the use of carbon nanotubes in polymer composites has attracted wide attention. Carbon nanotubes have a unique atomic structure, a very high aspect ratio, and extraordinary mechanical properties (strength and flexibility), making them ideal reinforcing compounds. Moreover, carbon nanotubes are susceptible to chemical functionalization, which broaden their applicability. For instance, surface functionalization of carbon nanotubes is an attractive route for increasing their compatibility with polymers in composites, also improving the dispersability in raw materials and the wettability.

Carbon nanotubes have shown a high potential to improve the mechanical properties of polymers, as well as the electrical properties [38–40]. Gojny et al. [41] reported that multiwalled carbon nanotubes could increase both tensile strength and fracture toughness of polymers. Flurian et al. [38] studied the influence of different carbon nanotubes on the tensile properties of polymers, as well as fracture toughness, and explained the contribution of nanomechanical mechanisms to enhancing fracture toughness.

Other studies of the effect of carbon nanotubes on the mechanical properties of polymers are calculated in Table 3.9.

3.3.1 MEDIUM-DENSITY POLYETHYLENE

Noroozu and Zeboria [57] have measured the effect of multiwalled carbon nanotubes on the mechanical and thermal properties of medium-density polyethylene matrix carbon nanotube nanocomposites.

TABLE 3.9

Effect of Carbon Nanotubes on Mechanical Properties

Polymer	Property Measured	References
Polyurethanes	Mechanical and rheological	42
Polyethylene composites	Morphological and rheological	43
Polyethylene glycol	Morphological and rheological	44
Polycaprolactone	Rheological	45
Nylon 10,10	Morphological and rheological	46
PET	Crystallinity	47
PC	Rheological	49
Polyetheramide and epoxy resins	Mechanical and electrical properties	50
Epoxy resins	Mechanical and electrical	51
PEEK	Rheological and electrical	52
PI	Crystallinity	53
Ethylene-vinyl acetate	Morphological and rheological	54
Polysilsesquioxane	Morphological and thermal	55, 56

PET = polyethylene terephthalate.

Incorporation of 0.5% to 5% of high-density polyethylene and multiwalled carbon nanotube composites improves Young's modulus from 0.4 to 0.56 GPa, yield stress from 16.34 to 19.02 MPa, and consumed energy up to 7% strain from 4729 to 5537 Nmm.

The interesting result found in the study is that the promotion of both the yield strength and Young's modulus of medium-density polyethylene due to the addition of multiwalled carbon nanotubes is much higher than that obtained via the conventional methods used by other investigators. For example, McNally [58] reported that adding multiwalled carbon nanotubes to polyethylene caused an increase of about 15% in yield stress and 22.5% in the Young's modulus. However, a 20% increase in yield stress and a 27.5% increase in the Young's modulus of medium-density polyethylene resulted from the incorporation of just 5% multiwalled carbon nanotubes. This observation proves that the distribution of multiwalled carbon nanotubes inside the matrix caused by milling is better than that achieved by conventional methods.

The following conclusions were reached in this work:

- The milling process can be a suitable method for producing medium-density polyethylene–multiwalled carbon nanotubes.
- Addition of carbon nanotubes to medium-density polyethylene causes a change in its morphology at constant milling parameters.
- The increases in both yield strength and Young's modulus of medium-density polyethylene due to the addition of medium-density polyethylene–multiwalled carbon nanotubes are much higher than those achieved by using conventional methods.

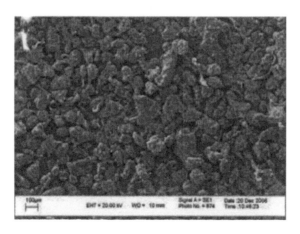

FIGURE 3.1 Electron micrograph of medium-density polyethylene before milling.

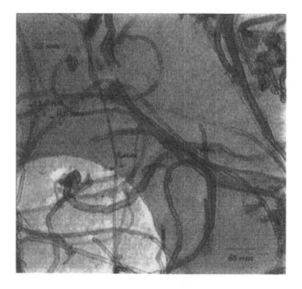

FIGURE 3.2 Transmission electron micrograph of carbon nanotubes.

Nanocomposites were examined by scanning electron microscopy and transmission electron microscopy.

Figures 3.1 and 3.2 show electron micrographs of medium-density polyethylene before milling and a transmission electron micrograph of carbon nanotubes, respectively.

Figure 3.3a–c show scanning electron micrographs of medium-density polyethylene, medium-density polyethylene containing 0.5% of carbon nanotubes, and medium-density polyethylene containing 1% carbon nanotubes after 10 h of milling. As is seen, at constant milling parameters (ball/powder ratio and milling time), addition of carbon nanotubes to medium-density polyethylene caused its morphology to change. This change occurred because the thermal conductivity of the carbon

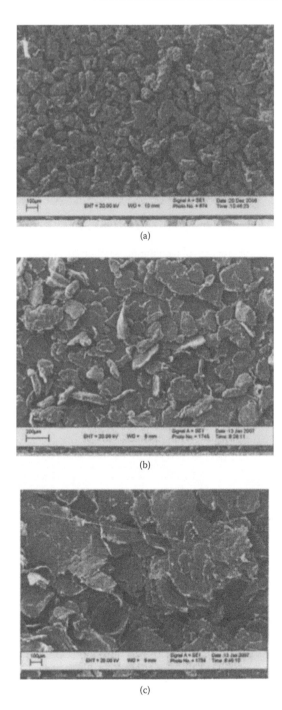

FIGURE 3.3   Scanning electron micrograph of (a) medium-density polyethylene, (b) medium-density polyethylene containing 0.5% of carbon nanotubes, and (c) medium-density polyethylene containing 1% of carbon nanotubes.

nanotubes was so much higher than that of the pure medium-density polyethylene, and this led to a dominant cold-weld mechanism during milling. In fact, the temperature experienced by the polyethylene particles during milling was very important in determining the nature of the final powder shape. It is reported that more than 90% of the mechanical energy imparted to the powders during milling is transformed into heat that raises the temperature of the powders. Increasing temperature makes polyethylene powders soft and finally melts them locally. On the other hand, the polyethylene powder particles can cold-weld together because of ball collisions. The presence of carbon nanotubes inside polyethylene powders leads to dispersion of the heat produced due to the larger milling area.

In fact, under the same milling conditions, the melted zone in polyethylene multiwalled carbon nanotube fiber is much bigger than that in the neat polyethylene.

3.3.2 Low-Density Polyethylene

Xiao et al. [59] carried out a detailed study of the mechanical and rheological properties of low-density polyethylene reinforced by the incorporation of multiwalled carbon nanotubes. It was found that the Young's modulus and tensile strength of the composites can increase by 89% and 56%, respectively, when the nanotube loading reaches 10 wt%. The curving and coiling of multiwalled carbon nanotubes play an important role in the enhancement of the composite modulus. It was also found that the materials experience a fluid–solid transition at the composition of 4.8 wt%, beyond which a continuous multiwalled carbon nanotube network forms throughout the matrix and in turn promotes the reinforcement of the multiwalled carbon nanotubes.

3.3.3 Ethylene–Propylene–Diene Terpolymers

Barroso-Bujans et al. [60] prepared nanocomposites of carbon nanotubes and sulfonated ethylene-propylene-norbornene terpolymer and compared the mechanical properties of carbon nanotubes and carbon black ethylene-propylene-norborene composites. They also evaluated the effect of carbon nanotube dispersion on the mechanical, thermal, and electrical behavior.

The curing process was analyzed by means of a rheometry test. The rheograms of ethylene-propylene-diene composites show that the maximum for a sample of sulfonated ethylene-propylene-norbornene was three times higher than that for virgin ethylene-propylene-norbornene terpolymer. This result suggests that the incorporation of sulfonate groups onto the carbon nanotube surface reduces the cross-limiting efficiency of the benzoyl peroxide catalyst used in the preparation.

It is well known that phenyl compounds interfere in the peroxide vulcanization by acting as radical traps and a consequent reduction in efficiency [61]. Nevertheless, it is worth noting that only a low percentage of benzene sulfonic groups (about 2%) is present in modified carbon nanotubes, which might indicate that other interactions are involved.

The effect of the pristine and modified ethylene-propylene-norbornene vulcanization reaction was also analyzed by differential scanning calorimetry under dynamic conditions at different heating rates (from 2 to 50°C/min), where the curing temperature increased as the heating rate increased. The activation energy (E_a) of the curing process was easily estimated by means of both the Ozawa and Kissinger equations:

$$E_a = 2.3R\{[d \log q/(1/T_p)]\} \tag{3.1}$$

$$E_a = 2.3R\{[d \log(q/T^2_p)]/[d(1/T_p)]\} \tag{3.2}$$

where R is the universal gas constant, q is the heating rate, and T_p is the temperature at the exothermic peak. It is evident that the lowest energy is required to vulcanize ethylene-propylene-diene norbornene rubber in the presence of pristine and modified carbon nanotubes, in particular, when the carbon nanotubes carry benzene sulfonic groups.

Incorporation of 5 phr of carbon nanotubes into ethylene-propylene-diene-norbornene leads to a slight increase in tensile strength coupled with a marked reduction in strain at break, indicating a decrease in polymer toughness and flexibility. This finding might be the result of a poor carbon nanotube dispersion, as indicated above, and may also be due to slight polymer–filler interactions. The carbon black has the expected reinforcing effect on the tensile strength of ethylene-propylene, which cannot yet be overcome by carbon nanotubes. Nevertheless, the sulfonated ethylene-propylene-norbornene, which showed poor mechanical properties, improved markedly both the tensile strength and the elongation at break with the introduction of carbon nanotubes. This is an important result since one of the disadvantages of sulfonated carbon nanotubes as an ion exchange membrane is its low mechanical resistance and flexibility. To explain this result, it is necessary to understand how the sulfonation affects the molecular weight and the cross-linking of ethylene-propylene-diene and why carbon nanotubes markedly improve the tensile strength in sulfonated ethylene-propylene-norbornene and do not in pristine ethylene-propylene-diene.

3.3.4 Epoxy Resins

Loos et al. [62, 63] studied the effect of carbon nanotube addition on the mechanical and thermal properties of epoxy components and matrices.

The effect the addition of small amounts of single-walled carbon nanotubes was discussed in terms of structural changes in the epoxy matrixes.

3.3.5 Polyamides

Sandler et al. [64] produced a series of reinforced polyamide 12 that were strengthened by the incorporation of multiwalled carbon nanotubes and carbon nanofibers. The dispersion and resulting mechanical properties for the nanotubes produced by electric arc were compared with those found by a variety of chemical vapor deposition techniques. A high quality of dispersion was achieved for the nanotubes, and the greatest improvements in stiffness were observed using aligned, substrate-grown carbon nanotubes. Entangled multiwalled carbon nanotubes led to the largest increase in yield stress. The degrees of polymer and nanofiller alignment and the morphologies of the alignment and the polymer matrix were assessed using x-ray diffraction and differential scanning calorimetry. The carbon nanotubes acted as nucleation sites under slow cooling conditions, but no significant variations in polymer morphology as a function of nanoscale filler type and loading fraction were observed. A simple rule-of-mixture evaluation of the nanocomposite stiffness revealed a higher effective modulus for the multiwalled carbon nanotubes than for the carbon nanofibers due to improved graphitic crystallinity. These workers compared effective nanotube modulus with those of nanoclays and common short glass and carbon fiber fillers in melt-blended polyamide composites. The intrinsic crystalline quality and the straightness of the embedded nanotubes were significant factors influencing the reinforcement capability. Wong et al. [65] prepared polyamide nanocomposites based on polyiminosebacoyl iminodecamethylene and multiwalled carbon nanotubes. Environmental scanning electron microscope micrographs of the fracture surfaces showed that there is not only an even dispersion of multiwalled carbon nanotubes throughout the polyamide matrix, but also a strongly interfacial adhesion with the matrix. The combined effect of more defects on multiwalled carbon nanotubes and low-temperature buckling fracture is mainly responsible for the broken tubes. Differential scanning calorimeter results showed that the multiwalled carbon nanotubes acted as a nucleation agent and increased the crystallization rate and decreased crystallite size. In the linear region, rheological measurements showed a distinct change in the frequency dependence of storage modulus, loss modulus, and complex viscosity, particularly at low frequencies. Wong et al. [65] conclude that the rheological percolation threshold might occur when the content of multiwalled carbon nanotubes is over 2 wt% in the composites.

3.3.6 POLYETHYLENE OXIDE

Jong et al. [44] prepared polyethylene oxide nanocomposites filled with functionalized multiwalled carbon nanotubes and characterized them using rheological and morphological measurements. This study investigated how the surface treatment of carbon nanotubes affects the carbon nanotubes' dispersion state. It is found that the nanocomposites have a higher effective volume fraction than the real volume fraction of the carbon nanotubes. The dispersion state of the carbon nanotubes was identified by using a field emission scanning electron spectroscope and transmission electron microscope. The rheological findings indicate that there exists a percolated network structure of the carbon nanotubes in the nanocomposites, which was confirmed by electrical conductivity measurements as well as by morphological observation.

3.3.7 MISCELLANEOUS POLYMERS

Studies on other polymers in which there have been incorporated carbon nanotubes include polycaprolation [45] and ethylene-vinyl acetate copolymers [54].

3.4 CARBON BLACK REINFORCEMENT

Carbon black has been used for the reinforcement of polymers, mainly for reinforcing elastomer and for UV protection, electromagnetic interference shielding, and antistatic shielding of polymers [66]. Novak et al. [67] studied the electroconductive high-density polyethylene-carbon black composite with improved toughness. Mechanical reinforcing effects of carbon black have not been widely studied.

Most of the research has been on the mechanical properties of nanocomposites at room temperature, and only a few researchers have studied the cryogenic properties of epoxy and its composite [68–70].

A study on the nanocomposite is important since it can affect the structural characteristics of a composite when it is used as a matrix of the laminates or the reinforcement of a foam core. It has been reported that the characteristics of the composite structure can be improved when a nanoclay-reinforced epoxy is used as a matrix of laminates. Antonio et al. [71] improved the damping coefficient and the energy dissipation characteristics of a glass-epoxy composite using nanoclay particles. Hosur et al. [72] improved the impact characteristic of a composite sandwich structure using the nanoclay infused foam.

3.4.1 EPOXY RESINS

Kim et al. [73] investigated the fracture toughness of nanoparticle-reinforced bisphenol A epoxy composites. In this study, ductile materials such as microsized rubber or polyamide particles were found to increase fracture toughness but decrease tensile strength. Carbon black also improved fracture toughness.

The addition of 2% carbon black bisphenol A epoxy resin increased its fracture toughness by about 18% from 0.73 to 0.86 MPa.

Uniform dispersion of the nanoparticles is essential. Putting the mixture through a three-roll mill five times was shown to be enough to get a reasonably uniform dispersion and structural fracture toughness. High shear mixing was also effective in dispersing carbon black particles in the viscous resins.

Flexural tests were performed by Kim et al. [73] to investigate the effect of the nanoparticles on the fracture toughness (Kt_c) of bisphenol A nanoparticle-reinforced epoxy composites at several low temperatures, including the cryogenic temperature of –150°C. Structure toughness values versus displacement curves of bisphenol A epoxy resins showed that the resins demonstrated ductile behavior at room temperature. However, at cryogenic temperatures (–150°C), it became brittle and the modulus and strength increased by 160% and 90%, respectively. In the case of steels, the fracture toughness decreases due to their increased brittleness as the environmental temperature decreases [74]. Though the epoxy became brittle, the fracture toughness increased significantly at the cryogenic temperature (–150°C). Nishijima et al. [75] reported that the epoxy with free space, which was defined as the unoccupied space within the molecules, could have high fracture toughness at cryogenic temperature since the free space still existed and the intermolecular forces between shrunk networks increased. The fracture toughness of the epoxy increased as the temperature was decreased, where the fracture toughness at –150°C was 2.3 times higher than that at room temperature.

Kim et al. [73] found that nanoparticles in the epoxy resin decreased the fracture toughness of the composite, although the fracture surface was much rougher than that of the epoxy, from which it might be concluded that the intermolecular forces between polymer networks of the epoxy were much more dominant than the toughness effect of the mixed nanoparticles at the cryogenic temperature.

If the nanoparticle-reinforced matrix is used for a composite laminate, the interlaminar fracture toughness and the impact characteristic can be improved at room temperature. However, at the cryogenic temperature, unreinforced epoxy may be able to provide better structural characterization.

In this work, the toughness effect of carbon black (Ketjenblack EC-300j, Ketjen Black International Co., Japan) and nanoclay (Cloisite 93A, Southern Clay Products, United States) on the modified bisphenol A–type epoxy resin (YD-114F, Kukdo Chemical, Korea) was investigated at room (25°C) and cryogenic (–150°C) temperatures.

At room temperature, a carbon black content of 3.0 wt% could increase the fracture toughness (Kt_c) value by 23% on average due to the toughening mechanisms of nanoscale crack branching and pinning effects.

At room temperature, 0.5% of nanoclay of content of only 0.5 wt% could increase the fracture toughness Kt_c value by 20%, while the nanoclay at 3.0 wt% could increased the Kt_c fracture toughness value by 50% on average, which was due to the tearing effect by the flake shape of nanoclay.

At the cryogenic temperature, the fracture toughness of the epoxy was 2.3 times higher than that at room temperature.

Fullerene consists of molecular balls made of 60 (termed C60) or more carbon atoms clusters linked together [74–76]. Each of the carbon atoms has two single bonds and one double bond that attach to other carbon atoms. This causes C60 to act more like a superalkene and superaromatic [74–76].

The antioxidant and reactive properties of fullerene make it suitable for use in rubbers, especially those exposed to UV lights as they become stronger and more elastic. Aging does not decrease mechanical properties of rubber and improves resistance to thermal degradation [77–80]. Filling the polymer with small amounts of fullerene could lead to a new set of properties that have specific advantages over those of rubbers currently in use.

Jurkowski et al. [81] found that the addition of fullerene to carbon black at a concentration between 0.065 and 0.75 phr increases Schob elasticity, hardness, and modulus of natural rubber. There is no substantial influence of fullerene on T_g, tan δ and G modulus, all evaluated by dynamic mechanical analysis at twisting within a temperature range of –150°C to 150°C (glassy state). At temperatures between 0°C and 150°C (rubbery state), an increase in modulus and some changes in the slope of segments in curves were observed. This could result from the additional strong physical junctions of the rubber network. This suggests a growth of degradation energies of the branching junctions and a related rise in the aging resistance as the concentration of fullerene increases. Simultaneously, it could be expected that some reduction occurs in tire temperature in service. Because of this, introduction of fullerene could be reasonable for tread rubbers and will reduce its price. Permittivity and dielectric loss angle are correlated with fullerene concentration. Compounding technology with raw rubber on available mechanisms could be easily implemented in the industry.

As stated above, fullerene increases Schob elasticity, hardness, and modulus at different elongations, for example, increasing fullerene content from 0.2 to 0.75 phr, shore elasticity from 6.5% to 77.1%, hardness from 31.8 to 34.7 shore A, and modulus at 200% from 1.43 to 2.43 MPa. This situation was expected by analogy with a case that carbon black causes an increase in the effective cross-link density of the rubber network. Accelerated aging increases rubber modulus, but the rate of modulus change is lower for higher fullerene contents.

The permanent set and temperature rise in the Goodrich test (Table 3.11) and the static hysteresis increase slightly or are independent of the concentration of fullerene; the static hysteresis temperature rise and permanent set in the Goodrich test for gum rubbers are much higher. However, the hardness lowers in comparison with those for the fullerene-containing rubbers. This is evidence that the physical branching junctions in the rubber network created by fullerene are stronger.

The rate of changes in modulus resulted from accelerated aging expressed by M_{agi}/M (Table 3.11). Those at 0.75 phr fullerene suggest that this filler substantially reduced the effects of rubber degradation.

3.4.2   Polyvinylidene Fluoride

Chen et al. [82] studied the mechanical and electrical properties of carbon black/polyvinylidene fluoride-tetrafluoroethylene-propylene films cross-linked with triethylene diamine.

This was under consideration as a possible binder material for lithium-ion battery electrodes.

Mechanical properties were studied as a function of the degree of cross-linking of this polymer with triethylene tetramine and bisphenol A. Bisphenol A cross-linking led to a distinct improvement in mechanical properties.

3.5   GRAPHITE-REINFORCED PLASTICS

Zheng et al. [84] and Choi and Kovac et al. [83] showed that graphite acted as a nucleating agent to induce the crystallization of high-density polyethylene and its composites.

However, the overall crystallinity and melting temperature decreased with increasing filler content. Tensile and dynamic mechanical analysis measurements indicated that expanded graphite was a better filler than untreated graphite. The overall improvement in mechanical properties was not great, but the mechanical strength and stiffness of high-density polyethylene were improved by the addition of the fillers.

The UK delegates take a leading role in international standards both technically and organizationally. For example, committee minutes and standard drafts need to be prepared in English as well as French and, within CEN only, German. In many cases, the original draft of a new test method will be in English, and considerable editing work is required to ensure a high-quality English text is produced. Subsequent translations can then be accurately undertaken, minimizing technical misinterpretation.

The work of the British Standard Institute (BSI) committees is now mainly devoted to inputs and ballot vote/comments on the ISO and CEN work programs. Some work is required to monitor/revise existing BS standards, but the majority will be replaced ultimately by CEN standards, as discussed within the CEN work program.

3.6 MONTMORILLONITE ORGANICALLY MODIFIED CLAY REINFORCEMENT

Various workers have investigated the mechanical and thermal properties of polymers in which organically modified clays and montmorillonites had been incorporated (Table 3.10). In general, it was found that increasing the clay contents of a polymer increased the storage and loss modulus as well as Young's modulus and reduced

TABLE 3.10
Influence of Incorporation of Organically Modified Clays and Montmorillonite Nanocomposites in Polymers

Polymer	Property Investigated	Reference
Montmorillonite		
Polyurethane	Glass transition (T_g), particle size, electrical	85
PVDF	Thermal	9
Polytrimethylene triphthalate	Heat stability, mechanical	10
Polyimide-amide	Rheological	86
Acrylonitrile-butadiene-styrene	Thermal	87
PE	Thermal	88
Polypropylene-ethylene-propylene-diene	Mechanical and viscoelastic	89
Hydroxy benzoic acid-hydroxy naphthoic acid copolymers	Thermal	90
LDPE	Rheological	91
Polyamide	Mechanical and rheological	92
Polystyrene nanocomposites	Thermal	93
Organically Modified Clays		
PLA	Rheological	94
PS	Rheological	95, 96
PE	Thermal	97
Ethylene-propylene-diene elastomers	Mechanical and rheological	98
PP	Mechanical and thermal, yield stress	99
Ethylene-vinyl acetate	Mechanical and thermal	100
Maleic anhydride–based PP composites	Rheological	101
Epoxies	Thermomechanical	102
Nylon 12	Thermal and rheological	103
High-impact PS-PET nanocomposites	Mechanical and rheological	104
Fluoroelastomers	Thermal	105
Polyacrylonitrile	Mechanical	107
Polyethylene oxide	Mechanical	108
Ethylene-propylene	Mechanical	106

LDPE = low-density polyethylene, PE = polyethylene, PLA = polylactic acid, PP = polypropylene, PS = polystyrene, PVDF = polyvinylidene fluoride.

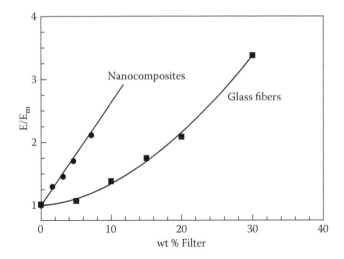

FIGURE 3.4 Comparison modulus reinforcement (relative to matrix polymer) increases for nanocomposites based on montmorillonite versus glass fiber (aspect ratio, ~20) for a nylon 6 matrix.

crystallinity. The glass transition temperature (T_g) increased and thermal stability tended to improve.

A common reason for adding fillers to polymers is to increase the modulus or stiffness via reinforcement mechanisms described by theories for composites [109–131]. Properly dispersed and aligned clay platelets have proved to be very effective for increasing stiffness. This is illustrated in Figure 3.4 by comparing the increase in the tensile modulus, E, of injection-molded composites based on polyamide 6, to the modulus of the neat polyamide matrix, E_m, for different volumes of percentage filler when the filler is an organoclay versus glass fiber. In this example, increasing the modulus by a factor of 2 relative to that of neat polyamide 6 requires approximately three times more mass of glass fibers than that of montmorillonite, platelets. Thus, the nanocomposite has a weight advantage over the conventional glass fiber composite. Furthermore, if the platelets are aligned in the plane of the sample, the same reinforcement should be seen in all directions within the plane, whereas fibers reinforce only along a single axis in the direction of their alignment [111]. In addition, the surface finish of the nanocomposite is much better than that of the glass fiber composite owing to the nanometer size of the clay platelets versus the 10–15 μ diameter of the glass fibers. A central question is whether the greater efficiency of the clay has anything to do with its nanometric dimensions, that is, a nanoeffect. All the experimental trends noted in this study can be explained using composite theory without invoking any nanoeffects.

Weiping et al. [133] and others [132, 134–137] have reported that nanoclay could increase fracture toughness on dry nanocomposites using the tensile and three-point bending methods.

Qui et al. [135] investigated the effect of several nanoclay additives that were mixed with epoxy resin using a mechanical stirrer. Properties measured included tensile modulus and fracture toughness. It was found that tensile strength and Vickers hardness both increased in clay reinforced with epoxy resin.

The beneficial effects of organically modified montmorillonite clay on the mechanical properties of various polymers are now discussed in further detail.

3.6.1 EPOXY REINS

Kim et al. [73] have investigated the effect of the addition of nanoclay on the strength and fracture toughness (Kt_c) epoxy resins (see Section 3.4.1).

When the nanoclay content was less than 1.5 wt%, the Kt_c value gradually increased. When the nanoclay content was 1.5 wt%, the fracture toughness was improved by 46% on average. The nanoclay is much more effective than carbon black. As little as 0.5 wt% of nanoclay could improve the fracture toughness by 10%. As the nanoclay content increased over 1.5 wt%, the average fracture toughness became saturated. The ductile–brittle fracture mode transition was observed in the range from 0.5 to 1.0 wt%. When the nanoclay content was 3.0 wt%, the maximum Kt_c value was achieved, which was 50% higher than that of the epoxy.

Kaya et al. [138] found that the incorporation of organically modified clay into epoxy resins improved mechanical properties of the nanocomposites.

Paul and Robeson et al. [139] have published an extremely informative review on the properties of exfoliated nanoclay-based nanocomposites. These have dominated the polymer literature, but there are a large number of other significant areas of current and emerging interest. This review details the technology involved with exfoliated clay-based nanocomposites and also includes other important areas, such as barrier properties, flammability resistance, biomedical applications, electrical/electronic/optoelectronic applications, and fuel cell interests. The important question of the nanoeffect of nanoparticles or fiber inclusion relative to their large-scale counterparts is addressed relative to crystallization and glass transition behavior. Other polymer (and composite)-based properties derive benefits from the nanoscale filler or fiber addition, and these questions are addressed.

3.6.2 ETHYLENE–PROPYLENE–DIENE TERPOLYMERS

Mirzazadeb and Katbal [140] prepared dynamically vulcanized ethylene-propylene-diene/propylene/clay nanocomposite thermoplastic vulcanizate. The nanocomposites were shown to have better mechanical properties than reference samples. The presence of nanoclay layers prevented chain crystallization, so the nanocomposites would have lower crystallinity. This could because of weak interface between two phases. The higher storage modulus propylene of nanocomposites was related to the effect of nanoclay layers on the ethylene-propylene-diene-propylene clay nanocomposite thermoplastic vulcanizate.

3.6.3 POLYAMIDES

A plot of the mechanical moduli of nylon 6 nanocomposites versus temperature shows that the intersection of these curves with the horizontal line is a good approximation to the heat distortion temperature of the materials [141]. This temperature, used as a benchmark for many applications, can be increased by approximately 100°C by addition of about 7% by weight of montmorillonite clay. This effect has been explained by simple reinforcement as predicted by composite theory without invoking any special nanoeffects. The effect of montmorillonite clay on the glass transition temperature of these materials is very slight [141]. Indeed, glass fibers cause an analogous increase in heat distortion temperature.

The addition of fillers, including clay, can also increase strength as well as modulus [142]; however, the opposite may also occur. The main issue is the level of adhesion of the filler to the matrix. For glass fiber composites, chemical bonding at the interface using silicone chemistry is used to achieve high-strength composites [143]. On the other hand, the modulus of glass fiber composites is not very much affected by the level of interfacial adhesion [143]. Unfortunately, at this time there is no effective way of measuring the level of adhesion of clay with polymer matrices. In addition, there are no methods of measuring chemical bonds between clay particles and polymer matrix analogs that are as effective as those used for glass fibers. Generally, addition of organoclays to ductile polymers increases the yield strength; however, for brittle matrices, failure strength is typically decreased [144–146].

Liu et al. [147] investigated nanocomposites composed of polyamide 6 and organoclay. These were studied by a range of techniques in order to determine the effect of processing conditions and clay concentration. The nanocomposites were produced by melt compounding from commercial samples. X-ray diffraction, transmission electron microscopy, optical microscopy dynamic mechanical analysis, and tensile tests were used to characterize the materials. Exfoliated nanocomposites were formed at clay concentrations below 5 wt%, and above this level mixed exfoliated and intercalated materials were produced. Smaller-scale unevenness in clay distribution was observed, as well as large-scale gradation across the injection-molded tensile bars. The crystal structure of the polyamide matrix was affected by the clay distribution. Nanocomposite samples were more brittle than neat polyamide 6, but showed higher Young's modulus. Yield strength was optimized at 5 wt% clay.

W.S. Chow and Z.A. Mohid-Isat [92, 148] measured tensile and other properties on nanocomposites of polyamide 6 and organically modified montmorillonite clay. Prepared by compounding followed by injection molding testing and dynamic mechanical thermal analysis (DMTA) were carried out. The effects of the organoclay on mechanical, morphological, and rheological properties of the nanocomposites are discussed in terms of exfoliation and dispersion of the organoclay in the polymer matrix.

Kusmon et al. [149] investigated the effect of maleated styrenemethylene-butylene-styrene on the mechanical and thermal properties of polyamide 6-polypropylene containing 4 phr of organically modified clay. With the exception of stiffness and strength, the addition of maleated styrene-ethylene-butylene-styrene into the polyamide 6 polypropylene-organo clay nanocomposites increased ductility, impact strength, and fracture toughness. The increase in ductility and fracture toughness at high testing speed could be attributed to the thermal blunting mechanism in front of the crack tip. Differential scanning calorimetry (DSC) results revealed that the presence of maleated ethylene-butylene-styrene had a negligible effect on the melting and crystallization behavior of the polyamide 6-polypropylene-organoclay nanocomposites. Thermogravimetric results showed that the incorporation of male-ated ethylene-butylene-styrene increased the thermal stability of the nanocomposites.

Using a nanoindentation technique, Shen et al. [150] studied the effects of clay concentrations on the mechanical properties (hardness, elastic modulus, and creep behavior) of exfoliated polyamide 6,6-clay nanocomposites. The results were discussed in conjunction with those obtained by dynamic mechanical analysis and optical microscopy, and also conjunction with changes in morphology, crystallinity, and x-ray diffraction.

3.6.4 POLYLACTIC ACID

Yingwei et al. [151] have investigated the rheological electrical and thermal properties of polylactic acid-organoclay nanocomposites. Both storage and loss modulus increased with clay loading at all frequencies.

The virgin polylactic acid and its clay nanocomposites, upon foaming with carbon dioxide and nitrogen as a blowing agent, showed, in the case of the nanocomposites, a reduction in all sizes and an increase in cell density.

3.6.5 POLYVINYLIDENE FLUORIDE

In general, it was found that increasing the content of organically modified clays and montmorillonite increased the storage and loss module and the Young's modulus of polyvinylidene fluoride, and also reduced crystallinity and increased the T_g. Also, the thermal stability tended to improve.

The storage modulus of the reinforced polymer was improved appreciably over the storage modulus obtained for the virgin polymer over the temperature range −100°C to 150°C.

3.6.6 MALEIC ANHYDRIDE–GRAFTED POLYPROPYLENE

Mirzazadeb et al. [152] prepared master batches of maleic anhydride–grafted polypropylene with different loadings of organoclay in an internal mixer and then added polypropylene to give polypropylene nanocomposites with different degrees of intercalation. The polypropylene nanocomposites prepared were melt-blended with ethylene-propylene-diene rubber in the presence of a vulcanizing system.

3.7 TALC REINFORCEMENT

Yua Sheng et al. [153] and Leong et al. [155] measured the tensile, flexural, and impact properties of ultra-fine-talc-filled polypropylene composites. Experimental results indicated that the addition of talc influenced the crystallization, morphology, and mechanical properties of talc-polypropylene composites. When the talc content was increased from 0% to 25%, the crystallinity of the polypropylene phase reached a maximum at 15% talc and the tensile and flexural strengths of talc-polypropylene composites also attained a maximum, while the impact strength declined to its lowest level. Scanning electron microscopy photographs of impact fracture surfaces indicated the crystal conformation of the polypropylene matrix and showed that the phase structure of composites varied with the content of talc in the talc-polypropylene composites.

Seldon [154] measured the weld line of injection-molded specimens of acrylo-nitrile-butadiene-styrene/polyphenylene oxide talc-filled polypropylene and glass fiber–reinforced polyamide 6. Fracture surfaces were examined by scanning electron microscopy. The effects of holding pressure, injection velocity, melt temperature, and mold temperature on weld line strength were studied using experimental design. For each parameter setting the weld line strength was measured and compared with the bulk strength via a weld line factor, defined as the strength of specimens with a weld line divided by the strength of specimens without a weld line. The highest weld line factors were obtained for unfilled materials molded using high melt temperature and holding pressure and low mold temperature.

Relative modulus versus talc clay-reinforced agent content for nanocomposites based on a thermoplastic polyolefin or a triphenylene oxide matrix polypropylene plus ethylene-based elastomer showed that relative to a particular filler content, an appreciably higher modulus content was obtained for the montmorillonite reinforc-ing agent than for talc [156]. Doubling the modulus of the phenylene oxide requires about four times more talc than montmorillonite, with the talc-reinforced polymer having an improved surface finish. In the case of the talc-reinforced polymer, exfoli-ation is appreciably better than with clay reinforcement. The talc-reinforced polymer has automotive applications.

3.8 SILICA REINFORCEMENT

3.8.1 Epoxy Resins

The incorporation of silica into epoxy resins has little effect on tensile strength. The tensile strength of reinforced epoxy resin is 30–84 MPa, compared to values of 68–72 MPa for silica-reinforced polymer. The flexural modulus falls from 80 GPa for the unreinforced polymer to 15 GPa for the reinforced polymerm while elonga-tion at break remains virtually unchanged.

Pasquel-Sanchez and Martinez [157] studied the effect on mechanical properties of incorporating silica nanoparticles into an epoxy resin based on the diglycidyl ether of bisphenol A.

3.8.2 Polypropylene

Choudhary et al. [158] determined the mechanical properties of polymers produced by the mixing of two incompatible components: nanoclay polypropylene and rice husk silica.

An analysis of the composite mechanical performance indicated that the state of mixing influenced the filler–matrix interaction and the composite tensile modulus. Results showed that an increase in composite crystallinity was related to the increase in tensile modulus. It was also found that reducing the particle size of the filler has

TABLE 3.11
Properties of Polymer-Silica Nanocomposites

Polymer	Property Studied	Reference
Ethylene diamine and maleic anhydride–grafted PP	Mechanical	161
Epoxies	Mechanical and thermal	157
PS	Viscoelastic	162
Polyacrylonitrile	Thermal	163
Poly(3,4-ethylene dioxythiophene)	Optical	164
Phenylene-vinylene oligomers	Optical	165
PP	Crystallinity	160
PET	Crystallinity	160
Ceramic-modified polyamide-imide	Mechanical and thermal	166
Polymethyl methacrylate	Thermal properties	167

a positive effect on the tensile modulus and the strength of the samples. Models proposed from statistical analysis were used to predict the composite tensile strength for the original particle size within the system parameter boundaries.

3.8.3 POLYESTERS

Hu et al. [159] studied the viscoelastic properties of silica-polymethyl methacrylate nanocomposites. These nanocomposites exhibited higher storage and loss modulus than those of pristine polymethyl methacrylate.

3.8.4 POLYETHYLENE TEREPHTHALATE

Chae and Kim [160] measured the effect of silica particles on storage modulus, loss tangent, and relaxation time of polyethylene terephthalate.

Other studies on polymer-silica nanocomposites are reviewed in Table 3.11.

3.9 CALCIUM CARBONATE REINFORCEMENT

3.9.1 POLYPROPYLENE

The incorporation of calcium carbonate into polymers reduces impact strength [168, 169], producing virtually no change in flexural modulus. The reported data for polypropylene are shown in Table 3.12.

3.9.2 POLYPROPYLENE VINYL ACETATE

Tang et al. [174] found that the tensile modulus of polypropylene-vinyl acetate copolymer increased with an increase of calcium carbonate filler weight fraction, and that the impact strength decreased rapidly when the weight fraction of calcium

TABLE 3.12

Mechanical Properties of Virgin and 20% Calcium Carbonate–Filled Polypropylene

Virgin	Virgin Polypropylene	Propylene Calcium Carbonate (20% Composite)
Tensile strength	29–38	2.6
Flexural modulus, MPa	1.5	2.10
Elongation at break, %	50–660	80
Izod compact strength, strength kj/m	0.07	0.05

carbonate fell below 10% and then decreased gradually with increasing weight fraction of calcium carbonate.

Liang [171] also observed this effect in the case of low-density polyethylene composites.

On the other hand, Xu et al. [172] found that addition of calcium carbonate led to a high tensile strength and modulus at 100% elongation, but a relatively low elongation at break. This is possibly a result of the evolution of molecular interactions and the formation of additional physical cross-link induced by the filler network. Addition of the filler increased the glass transition temperature of polydimethyl siloxane. The width of the linear viscoelastic region in which dynamic storage modulus is constant at low strain amplitudes narrowed as the amount of filler increased. A characteristic plateau phenomenon appeared in low-frequency regions: the width and height of the modulus plateau increased. This phenomenon was also ascribed to the formation of a filler network as a result of filler–polymer and filler–filler interactions.

3.9.3 POLYETHER ETHER KETONE

Zhou et al. [173] studied the effects of surface treatment of calcium carbonate particles with sulfonated polyether ether ketone on the mechanical and thermal properties of composites with polyether ether ketone in various proportions prepared using a twin-screw extruder. These workers used tensile, impact, and flexural testing, thermogravimetric analysis, differential scanning calorimetry, and scanning electron microscopy. The influences of filler particle, loading, and surface treatment on deformation and crystallinity of polyether ether ketone were discussed.

Calcium carbonate has also been used as a reinforcing agent for a range of polymers, including polyether ether ketone [174], acrylonitrile-butadiene-styrene terephthalate [175], polypropylene, and polydimethyl siloxane.

3.10 MISCELLANEOUS REINFORCING AGENTS

Other materials that have been incorporated into polymers to modify mechanical and other properties include calcium sulfate in styrene-butadiene rubber [174–178], barium sulfate in polyethylene [179], barium sulfate in polypropylene [174], aluminum in epoxy resins [178], kaolinite-muscovite in polyvinyl chloride-polybutyl acrylate

TABLE 3.13

Effect of Aluminum Powder and Mineral Fillers on Mechanical Properties of Three Polymers

Polymer	Epoxy Resin		Polyacetal		Polyallylophthalate	
	uf	f	uf	f	uf	f
Aluminum Powder Filler						
Tensile strength, MPa	30–84	48	73	55	—	—
Flexural modulus, GPa	3.5	9.5	2.6	2.3	10.6	9.5
Elongation at break, %	1.3	1.0	65	50	—	—
Impact strength, kj/m	—	—	—	—	—	—
Mineral Filler						
Tensile strength, MPa	34–84	58	—	—	70	57
Flexural modulus, GPa	3.5	1.1	—	—	10.6	9.5
Elongation at break, %	1.3	1.8	—	—	0.90	1.9
Impact strength, kj/m					0.40	0.38

uf = unfilled polymer, f = polymer fillers.

[179], aluminum in polyethylene [180], nickel in epoxy resins [181], copper in polyethylene [182], gold in polystyrene [183], caesium bromide in polyvinyl alcohol [184], nickel-cobalt-zinc ferrite in natural rubber [185], and basalt in polycarbonate [186, 187].

The limited amount of information available on the mechanical properties of polymers filled with two of these agents, namely, aluminum powder and minerals, is reviewed in Table 3.13.

3.11   NATURAL BIOMATERIAL REINFORCEMENT

An early example of the use of biomaterials to reinforce is the incorporation of straw or horsehair to improve the strength of bricks and mortar.

Natural materials that have been used to modify polymer properties include wood sawdust [188, 189], coconut fiber [167], carbon fibers [37], oat husks, cocoa, shells [190], sugarcane fibers [191], and banana fiber [193]. Monhonty and coworkers [193] observed a 70% increase in the flexural strength of polypropylene to which has been added 30% jute with a maleic anhydride–grafted polypropylene coupling agent.

3.12   ORGANIC REINFORCING AGENTS

Lin et al. [194] found that reinforcement of polyurethane elastomer with wholly rigid aromatic m-phenylene isophthalamide block copolymers results in glass transition temperatures below 0%. Such block copolymers have improved the reinforcing effect, which is reflected in both their tensile strength and elongation when compared with virgin polyurethane.

Evrard et al. [195] evaluated powdered polyamides of different types as semi-reinforcing fillers in carboxylated nitrile rubber. They evaluated the filler vulcanizates for tensile properties, tear strength, coefficient of friction, filler matrix adhesion, and swelling in various fluids.

Dynamic mechanical analysis has also been used to determine the mechanical and thermal properties of low-density polyethylene and ethylene-propylene-diene terpolymer containing jute filler, which had improved flexural and impact properties compared to those of the base polymer [198]. Jeong and coworkers [196] and others [195] investigated the dynamic mechanical properties of a series of polyhexamethylene terephthalate, poly(1,4-cyclohexylenedimethylene terephthalate), and random copolymers thereof in the amorphous state as a function of temperature and frequency. The effect of copolymer composition on dynamic mechanical properties was examined and the dynamic mechanical properties interpreted in terms of the cooperativity of segmental motions.

3.13   METHODS FOR REINFORCED PLASTICS

Sims et al. [197] has reviewed some UK contributions to the development of international specification and test method standards for fiber-reinforced plastics. Sims et al. state that the UK has been active in international standardization for many years and has produced, in BS 4994 and BS 6464 for GRP pressure vessels and pipes, two of the few readily accessible design standards for composites. The UK works with the International Organization for Standardization (ISO) and the Comité Européen de Normalisation (CEN). This is happening at a time when there is a demand for increased testing efficiency with concurrent reductions in cost. For composites, there is a need to harmonize and then validate the many existing versions of established test methods. In the longer term, the aim should be to reduce the timescale for standardization. Dr. Sims et al. illustrates in this paper how the UK, through organizations such as the British Plastics Federations and the National Physical Laboratory, is taking a major role in research, harmonization, drafting, and validation of international standards.

Sims et al. [197] have pointed out that a redraft of ISO 527 tensile testing for plastics has resulted in two complementary parts for composites: Part 4 based on ISO 3268 [199] covers isotropic and orthotropic materials, and Part 5 covers unidirectional materials. There is a need to harmonize these two methods into a new Part 5. Inputs were included from ISO, ASTM, CRAG, and EN aerospace methods. The remaining three parts are: Part 1—general principles, Part 2—plastics and molding materials (including short fiber–reinforced materials), and Part 3—films. ISO DIS 527-5 has in parallel been voted as a CEN standard (EN 527-5) and should, with Part 4, replace EN61 [212] and possibly two aerospace EN standards [34, 42, 132, 160, 163–165, 199–213].

Obtaining agreement between ISO, CEN general series, EN aerospace–carbon only, ASTM, and JIS in specimen sizes for unidirectional reinforced materials was a notable first step toward harmonization of the test methods, particularly because the 1.25-cm-wide ASTM specimen [215] has been in extensive use for many years. The agreed dimensions are 15×1 mm for $0°$ and 25×2 mm for $90°$ fiber direction specimens. After agreement on the specimen sizes (which is the same in all standards series, except EN aerospace for glass fiber systems [214]), several other aspects of the test method were decided upon during drafting of the new standard.

A complete list of recommended mechanical test methods for fiber-reinforced plastics has been reported [216–245].

CEN is developing a new series of standards in support of the European single market and EC directives. Technical committees are working at most levels, from major installations to material specification and testing. This is occurring at a time when there is a demand for increased testing efficiency with concurrent reductions in the associated cost. For composite materials, there is initially a need to harmonize and then validate the many existing versions of established test methods. In the longer term, research on test methods as supported by the UK Department of Trade and Industry and that undertaken under the Versailles Agreement and Advanced Materials and Standards (VAMAS) international prestandards program should aim to reduce the timescale for standardization. Validation of test methods, which, due to the legal position of CEN standards within the single market, is particularly essential, must be considered at an early stage. The UK, through organizations such as the British Plastics Federation and the National Physical Laboratory, is taking on a major role in research, harmonization, drafting, and validation of international standards.

The main forums for standardization in the UK were for many years either through BSI itself or via its role in support of ISO standards. For the materials testing and specification level, the work within ISO is undertaken in Technical Committee (TC) 61 (Plastics) mainly within Subcommittee SC13 (Composites and Reinforcement Fibers). In addition to this, many other subcommittees, such as SC2 for dealing with mechanical test methods, SC4 for fire tests, and SC12 for thermosets, develop relevant test methods covering both unreinforced and reinforced plastics.

The subcommittees forming SC13 are shown in Table 3.14. It should be noted that the work program on test methods for laminates is undertaken in Working Groups (WGs) 14 and 16.

TABLE 3.14
Structure of ISO TC61/SC13

Number	Title	Country
SC13	Composites and reinforcement fibers	Gullermin/France
SC13/WG11	General standards	Downey/UK
SC13/WG12	Reinforcement fibers	Charlier/Belgium
SC13/WG13	Glass fiber standards	Toustou/France
SC13/WG14	Carbon fiber standards	Matsui/Japan
SC13/WG15	Prepregs and molding compounds	Garcia-Folques/Spain
SC13/WG16	Composites mechanical test methods	Krawczak/France

The BSI committees related to these ISO committees are

PRI/41	Fibers for reinforcement and test methods for composites (re: SC13)	
PRI/40	Thermosets (re: SC12 and SC13)	
PRI/26	Burning behavior (re: SC4)	

There are currently in excess of 65 standards related to composites at the material level. The majority of these are published standards, mainly requiring review and maintenance, with about 10%–20% as new work items. Although many ISO standards have been dual numbered as BS standards, it has also been possible to produce national standards of the same scope better suited to UK requirements or local historical practice. In addition, there are some 22 BS standards, which are not dual numbered. Similarly, the United States has concentrated on ASTM standards produced by the D-30 committee. This committee has had little involvement in ISO, in spite of its background or relevant test methods.

REFERENCES

1. L. Huang, Q. Yuan, W. Jiang, L. An, and S. X. Jiang, *Journal of Applied Polymer Science*, 2004, 94, 1885.
2. A.P. Gupta, U.K. Saroop, G.S. Jhe, and M. Verme, *Polymer Plastics Technology and Engineering*, 2003, 42, 297.
3. A.P. Gupta, Union Carbide Chemie Co., Inc. recycle Citation 23, November, February 1993, p. 17.
4. P.J. Farrington and R.G. Speight, *Proceedings of the 63rd SPC Annual Conference (ANTEC2005)*, Boston, May 1–5, 2005, pp. 102, 318.
5. MS Union, *Additives Can Improve Properties of Recycled Glass Reinforced Plastics*, 155N 0048-7457, Carbide Refuse/Recycle Citation 23, no. 2, February 1993, p. 12.
6. H. Eriksson, *Kunststoffe*, 2009, 9, 88.
7. A.P. Gupta, U.K. Saroop, and M. Verme, *Polymer Plastics Technology and Engineering*, 2004, 48, 937.
8. S. Mistra and N.G. Shimpi, *Journal of Polymer Science*, 2007, 104, 2018.
9. Y.P. Wang, X. Goa, R.M. Wong, H.G. Liu, C. Yang, and Y.B. Xiong, *Reactive and Functional Polymers*, 2008, 687, 1170.
10. Z. Liu, K. Chen, and D. Yan, *Polymer Testing*, 2004, 23, 323.
11. T. Mikolajczyk and M. Olenjik, *Fillers and Textiles in Eastern Europe*, 2007, 15, 26.
12. L.N.P. Engineering Plastics, *Plastics and Rubbers Weekly*, 1997, no. 1698, 13.
13. R. Khoshravan and C. Bathias, Delamination tests, UDEB of ENF composite specimens, annual report, VAMAS W95, 1990.
14. M.Y. Keating, L.B. Malone, and W.D. Saunders, *Journal of Thermal Analysis and Calorimetry*, 2002, 69, 37.
15. M. Koh, B. Larsen, T. Mackey, and F. Tafan, SPEC 2002, *Proceedings of Plastics Impact on the Environment*, Detroit, February 13–14, 2002, p. 55.
16. L. Huang, Q. Yuan, W. Jiong, and L.A.S. Jiang, *Journal of Applied Polymer Science*, 2004, 94, 1885.
17. J.L. Lyons, *Polymer Testing*, 1998, 17, 237.
18. G.P. Karayannidis, D.N. Bikiaris, and G.Z. Papageorgiou, private communication.
19. ASTM D430-95, Standard test methods for rubber deterioration, dynamic fatigue, 2000.
20. T. Miyata and R. Yomaokn, *Kobushi Ronbunshu*, 2002, 59, 415.
21. F.H. Gojny, M.M.G. Wichman, B. Fiedler, W. Bauhofer, and K. Schulte, *Composites, Part A: Applied Science and Manufacturing*, 2005, 11, 1525.
22. Y. Nischitani, I. Sekiguchi, B. Hausnerova, N. Zolpazilova, and T. Kitono, *Polymers and Polymer Composites*, 2007, 15, 111.
23. M.J. Gehrig, S. Kelly, and M. Carr, May 2006, available from www.ptonline.com.
24. R. Seldon, *Polymer Engineering and Science*, 1997, 37, 205.

25. G.D. Sims, *International Standardisation of Coupon and Structural Element Test Methods*, ECCM Composites: Testing and Standardization 2, Hamburg, Germany, 1994.
26. C.-H. Chen and K.-C. Lein, *Polymers and Polymer Composites*, 2006, 14, 155.
27. L. Priya and J.P. Jog, *Journal of Polymer Science, Part B: Polymer Physics*, 2003, 41, 31.
28. Q. Yan and W. Jiang, *Polymers for Advanced Techniques*, 2004, 15, 409.
29. Y. Chen, J.F. Davaios, and I. Ray, *Journal of Composites for Construction*, 2006, 279.
30. W. Leong, A.M. Ishak, and A. Ariffin, *Journal of Applied Polymer Science*, 2004, 915, 3327.
31. J.L. Abot, A. Yasmin, A.J. Jacobson, and I.M. Daniel, *Composite Science and Technology*, 2004, 64, 263.
32. G. Sudbury, *Journal of Advanced Material*, 2010, 42, 28.
33. B. Hausrenova, N. Zdrazilova, and T. Kitano, *Saha Polimary*, 2006, 51, 33.
34. S. Vidhate, E. Ogunsona, J. Chung, and N.A. D'Souza, *Proceedings of the 66th SPE Annual Technical Conference*, Milwaukee, WI, 2008, p. 74.
35. R.D. Patton, C.U. Pittman, L. Wang, J.R. Hill, and A. Day, *Composites, Part A: Applied Science of Manufacturers*, 2002, 33A, no. 2, 243.
36. C.Z. Lai, L. Christiansen, and S.R. Dahn, *Journal of Applied Polymer Science*, 2004, 91, 2949.
37. E.M. Wouterson, F.C.Y. Boey, X. Ha, and S.C. Wong, *Polymer*, 2007, 48, 3183.
38. H.G. Fioran, H.G. Malte, F. Bodo, and S. Karl, *Composites Science and Technology*, 2005, 65, 2300.
39. J. Sandler, M.S.P. Shaffer, T. Prasse, W. Bauhofer, K. Shulte, and K. Windle, *Polymer*, 1999, 40, 5967.
40. C.A. Martin, J.K.W. Sandler, M.S.P. Shaffer, M.K. Scharz, W. Bauhoger, K. Schulte, et al., *Composites Science and Technology*, 2004, 64, 2309.
41. F.H. Gojny, M.H.G. Wichman, U. Köpke, B. Fiedler, and K. Schulte, *Composites Science and Technology*, 2004, 64, 2363.
42. H.-C. Kuan, C.-C.M. Ma, W.-P. Chong, S.-M. Yuen, H.-H. Win, and T.-M. Lee, *Composites Science and Technology*, 2005, 65, 1703.
43. K.Q. Xiao, L.C. Zhang, and I. Zarudi, *Composites Science and Technology*, 2007, 67, 177.
44. Y.S. Song, Polymer Engineering and Science, 2006, 46, 1350.
45. D. Wu, L. Wu, Y. Sun, and M. Zhang, *Journal of Polymer Science, Part B: Polymer Physics*, 2007, 45, 3137.
46. B. Wang, G. Sun, and J. Lui, *Polymer Engineering and Science*, 2007, 47, 1610.
47. S. Tzavalas, D.E. Mouzakis, V. Drakonakis, and V.E. Gregoriou, *Journal of Polymer Science, Part B: Polymer Physics*, 2008, 46, 668.
48. M. Lai, J. Li, J. Yang, J. Liu, X. Tong, and H. Cheng, *Polymer International*, 2006, 53, 1479.
49. Y.T. Sung, M.S. Han, K.H. Song, J.W. Jung, H.S. Lee, C.K. Kum, J. Loo, and W.N. Kim, *Polymer*, 2006, 47, 4434.
50. T. Aoki, Y. Ono, and T. Ogasawara, *Proceedings of the American Society for Composites 21st Technical Conference*, Dearborn, MI, 2006, paper 31.
51. S.-M. Yeun, C.-C.M. Ma, C.-Y. Cuang, Y.-H. Hsoao, C.L. Chiang, and A.-D. Yu, *Composites, Part A: Applied Science of Manufacturers*, 2008, 39, 119.
52. D.S. Bangarusam Path, V. Alstradt, H. Ruckdeseschel, J.K.W. Sandler, and M.S.P. Shaffer, *Journal of Plastics Technology*, 2008, 6, 19.
53. V.E. Yudin, V.M. Scetlidnya, A.N. Shumkaov, D.G. Letenko, A.Y. Feldman, and G. Marom, *Micromolecular Rapid Communications*, 2005, 26, 885.
54. S. Peterbroveck, M. Alexandre, J.B. Nagy, N. Morean, A. Destrec, F. Monteverde, A. Pulmont, R.J. Jreome, and P. Dubois, *Macromolecular Symposia*, 2005, 22, 115.

55. S.-M. Yuen, C.-C.M. Ma, C.-C. Teng, H.-H. Wu, H.-C. Kuan, and C.-L. Chiang, *Journal of Polymer Science, Part B: Polymer Physics*, 2008, 46, 472.
56. H.C. Kuan and C.C. Ma, *Composites Technologies for 2020, Proceedings of the Fourth Asian Australasian Conference on Composite Materials (AACM-4)*, Sydney, Australia, June 6–9, 2004, p. 736.
57. M. Noroozi and S.M. Zaboria, *Journal of Vinyl and Additive Technology*, 2010, 147, 57.
58. T. McNally, *Journal of Polymers*, 2005, 46, 8222.
59. K. Xiao, L.E. Zhang, and I. Zarudi, *Composites Science and Technology*, 2007, 67, 177.
60. F. Barroso-Bujans, M. Arroyo, E. San Juan, I. Rodrigues-Romos, M. Pelez-Cabero, E. Riand, and M.A. Lopez-Manchado, Institute de Ciencic Technologia de Polimeros, Juan de la Cierva, 3, 28006, Madrid, Spain.
61. J.B. Class, *Rubber World*, 1999, 220, 35.
62. M.R. Loos, S.H. Pezzin, S.C. Amico, C.P. Bergmann, and L.A.F. Coelho, *Journal of Materials Science*, 2008, 18, 6064.
63. M.R. Loos, L.A.F. Coelho, S.H. Pezzin, and S.C. Amico, *Journal of Materials Research*, 2008, 11, 347.
64. J.K.W. Sandler, S. Pagel, M. Cadek, F. Gogny, M. Vants, J. Lachman, W.J. Blau, and M.S.P. Schaffer, *Polymer*, 2004, 45, 2001.
65. B. Wong, G. Gun, X. Le, and J. Lin, *Polymer Engineering and Science*, 2007, 47, 610.
66. J.B. Donnet, R.C. Bansal, and M.J. Wang, *Carbon Black Science and Technology*, Marcel Dekker, New York, 1993.
67. I. Novak, I. Krupa, and I. Janigova, *Carbon*, 2005, 43, 841.
68. F. Sawa, S. Nishijima, and T. Okada, *Cryogenics*, 1995, 35, 767.
69. S. Nishijima, Y. Honda, and T. Okada, *Cryogenics*, 1995, 35, 779.
70. C.J. Huang, S.Y. Fu, Y.H. Zhang, B. Lauke, L.F. Li, and L. Ye, *Cryogenics*, 2005, 45, 450.
71. F.A. Antonio, V.D. Làzaro, and V.D. Horàcio, *Composite Structures*, 2008, 83, 324.
72. M.V. Hosur, A.A. Mohammed, S. Zainuddin, and S. Jeelani, *Composite Structures*, 2008, 82, 101.
73. B.C. Kim, S.W. Park, and D.G. Gil, *Composites Structures*, 2008, 86, 69.
74. T.L. Anderson, *Fracture Mechanics: Fundamentals and Applications*, CRC Press, Boca Raton, FL, 1995.
75. S. Nishijima, Y. Honda, and T. Okada, *Cryogenics*, 1995, 35, 779.
76. P.R. Birkett, P.B. Hitchcock, H.W. Kroto, R. Taylor, and D.M.R. Walton, *Nature*, 1992, 357, 479.
77. G.R. Hamed, *Rubber Chemistry and Technology*, 2000, 73, 524.
78. Russian Patent 2151,781, Institute of the Problems of Chemical Physics, Russian Academy of Sciences, 2000.
79. F. Cataldo, *Fullerene, Nanotubes, and Carbon Nanostructures*, 2001, 9, 497.
80. F. Cataldo, *Fullerene, Nanotubes and Carbon Nanostructures*, 2001, 9, 515.
81. B. Jurkoweska, B. Jurkowowkii, P. Kaurowski, S.S. Resetkii, V.N. Koval, L.S. Pincuk, and Y.A. Olkhov, *Journal of Applied Polymer Science*, 2006, 100, 390.
82. Z. Chen, L. Christensen, and J.R. Dahl, *Journal of Applied Polymer Science*, 2004, 5, 2949.
83. J. Choi and S.Y. Kavak, *Macromolecules*, 2004, 37, 3745.
84. W. Zheng, X. Lu, and S.C. Wong, *Journal of Applied Polymer Science*, 2004, 91, 2787.
85. H.-T. Lee and L.-H. Lin, *Macromolecules*, 2006, 39, 6133.
86. T. Mikolajczyk and M. Olejnik, *Fillers and Textiles in Eastern Europe*, 2007, 15, 26.
87. B. Pourabbas and H. Azimi, *Journal of Composite Materials*, 2008, 42, 2499.
88. K. Stoeffler, P.G. Lafleur, and J. Denault, *Polymer Engineering and Science*, 2008, 48, 1449.
89. H. Mirzasadeh and A.G. Katbaly, *Polymers for Advanced Technologies*, 2006, 17, 975.

90. S. Banydopadhyaly, S.S. Ray, and M. Bousina, *Macromolecular Chemistry and Physics*, 2007, 208, 1979.
91. Y.C. Kim, S.J. Lee, J.C. Kim, and H. Co, *Polymer Journal* (Japan), 2005, 37, 206.
92. W.S. Chow and Z.A. Mohd Isak, *Express Polymer Letters*, 2007, 1, 77.
93. D.-R. Yei, S.-W. Kuo, H.-K. Fe, and F.-C. Chang, *Polymer*, 2005, 46, 741.
94. D.Y. Wei, S. Innace, E. Di Maio, and L. Nicolais, *Journal of Polymer Science, Part B: Polymer Physics*, 2005, 43, 689.
95. S. Tonoue, L.A. Utracki, A. Garcia Rejon, P. Sammut, I. Pesneau, M.R. Kamal, and I. Layngaae-Jorgensen, *Polymer Engineering and Science*, 2004, 44, 1061.
96. K. Stoeffler, P.G. Lafleur, and J. Renault, *Proceedings of the 64th SPE Annual Conference (ANTEC 2006)*, Charlotte, NC, 2006, p. 263.
97. J. Strange, S. Waechtex, H. Muenstedt, and H. Kaspar, *Macromolecules*, 2007, 40, 2409.
98. J. Jachowicz, R. McMuller, C. Wu, L. Senak, and D. Koelmel, *Journal of Applied Polymer Science*, 2007, 105, 190.
99. F. Yang, X.-H. Xiong, Q. Wang, and L. Li, *Polymer Materials Science and Engineering*, 2007, 23, 190.
100. A. Lindner, B. Lestriez, S. Mariot, C. Creton, T. Maevis, B. Luehmann, and R. Brummer, *Journal of Adhesion*, 2006, 82, 267.
101. L. Andreozzi, V. Castelvetro, M. Faetti, M. Giordano, and F. Zulli, *Macromolecules*, 2006, 39, 1880.
102. J. Chan, X. Liu, and D. Yang, *Journal of Applied Polymer Science*, 2006, 102, 4040.
103. J.L. Zhang, L. Shen, W.C. Li, and Q. Zheng, *Journal of Material Science and Engineering*, 2006, 22, 123.
104. K.Y. Kim, J. Nam, and S.M. Lee, *Journal of Applied Polymer Science*, 1006, 99, 2132.
105. P. Prasopnatre, S. Sarouc, and C. Sirisinha, *Journal of Applied Polymer Science*, 2009, 111, 1051.
106. H.W. Wong, I.T.C. Chang, J.M. Yeh, and S.J. Lion, *Journal of Applied Polymer Science*, 2004, 91, 1368.
107. Q. Zhang, H. Yang, and Q. Fu, *Polymer*, 2004, 45, 1913.
108. K.F. Stranbecker and E. Manias, *Chemistry of Materials*, 2003, 15, 844.
109. J.W. Liu, Y.J. Wang, and Z. Sun, *Journal of Applied Polymer Science*, 2004, 92, 485.
110. T. Faras and D.R. Paul, *Polymer*, 2003, 44, 4993.
111. K.Y. Lee and D.R. Paul, *Polymer*, 2005, 46, 9064.
112. K.Y. Lee, K.H. Kim, S.K. Jeoung, S.I. Ju, J.H. Shim, N.H. Kim, et al., *Polymer*, 2007, 48, 4174.
113. T. Mori and K. Tanaka, *Acta Metallurgica*, 1973, 21, 571.
114. J.C. Halpin and L. Kardos, *Polymer Engineering and Science*, 1976, 16, 344.
115. T.S. Chow, *Journal of Polymer Science*, 1978, 16, 959.
116. T.S. Chow, *Journal of Polymer Science*, 1978, 16, 967.
117. T.S. Chow, *Journal of Materials Science*, 1980, 15, 1873.
118. P.J. Hine, H.R. Lusti, and A.A. Gusev, *Composite Science and Technology*, 2002, 62, 1445.
119. H.R. Lusti, P.J. Hine, and A.A. Gusev, *Composite Science and Technology*, 2002, 62, 1927.
120. M. Van Es, F. Xiqiao, J. van Turnhoult, and E. van der Giessen, in S. Al-Malaika, A. Golovoy, and C.A. Wilkie (eds.), *Speciality Polymer Additives, Principles and Applications*, Blackwell Science, Oxford, 2001, p. 391.
121. D.A. Brane and J. Bicerano, *Polymer*, 2002, 43, 369.
122. J.-J. Luo and I.M. Daniel, *Composite Science and Technology*, 2003, 63, 1607.
123. L. Zhu and K.A. Narh, *Journal of Polymer Science, Part B: Polymer Physics*, 2004, 42, 2391.
124. J. Tsai and C.T. Sun, *Journal of Composite Materials*, 2004, 38, 567.
125. M.A. Sharaf and J.E. Mark, *Polymer*, 2004, 45, 3943.
126. J. Wang and R. Pyrz, *Composite Science and Technology*, 2004, 64, 925.

127. J. Wang and R. Pyrz, *Composite Science and Technology*, 2004, 64, 935.
128. P.D. Shepard, F.J. Golemba, and F.W. Maine, *Advances in Chemistry Series*, 1973, 134, 41.
129. N. Sheng, M.C. Bouce, D.M. Parks, G.C. Rutledge, J.I. Abes, and R.E. Cohen, *Polymer*, 2004, 45, 487.
130. K. Hbaieb, Q.X, Wang, Y.H.J. Chia, and B. Cotterell, *Polymer*, 2007, 48, 901.
131. S. Sen, J.D. Thomin, S.K. Kumar, and P. Keblinski, *Macromolecules*, 2007, 40, 4059.
132. A. Usiki, Y. Kojima, M. Kawasumi, A. Okada, Y. Fukushima, and T. Kuranchi, *Journal of Materials Research*, 1993, 81, 179.
133. L. Weiping, V.H. Suong, and P. Martin, *Composite Science and Technology*, 2005, 65, 2364.
134. W. Lei, W. Ke, C. Lind, Z. Yongwei, and H. Chaobin, *Composites, Part A: Applied Science of Manufacturers*, 2006, 37, 1890.
135. B. Qi, Q.X. Zhaing, M. Bannister, and Y.W. Mai, *Composite Structures*, 2006, 75, 514.
136. M.W. Ho, C.K. Lam, K.T. Lau, H.L. Dickson, and H. David, *Composite Structures*, 2006, 75, 415.
137. I. Reade, *Asian Plastics News*, January/February 2001 p. 24.
138. F. Kaya, M. Tanoglu, and S. Okur, *Journal of Applied Polymer Science*, 2008, 109, 834.
139. D.R. Paul and L.M. Robeson, *Polymer*, 2008, 49, 3187.
140. H. Mirazabdeh and A.A. Katbal, *Polymers for Advanced Technologies*, 2006, 11, 975.
141. T.D. Forne and D.R. Paul, *Polymer*, 2003, 44, 4993.
142. T.D. Forne, P.J. Yion, M. Keskkula, and D.R. Paul, *Polymer*, 2001, 42, 9929.
143. D.Y. Laura, H. Kekkula, J.E. Barlow, and D.R. Paul, *Polymer*, 2002, 43, 4673.
144. H.A. Stretz, D.R. Paul, R. Li, H. Keskkula, and P.E. Cassidy, *Polymer*, 2005, 46, 2621.
145. R.K. Shah, D.L. Hunter, and D.R. Paul, *Polymer*, 2005, 46, 2646.
146. H.A. Stretz, D.R. Paul, and P.T. Cassidy, *Polymer*, 2005, 46, 3818.
147. T. Liu, W.C. Tiju, C. He, S.S. Na, and T.S. Chuny, *Polymer International*, 2004, 53, 392.
148. W.S. Chow and Z.A. Mohid-Isat, *Express Polymer Letters*, 2007, 1, 77.
149. A.U. Kusmon, M. Ishak, W.S. Chow, and T.T. Rochmad, *European Polymer Journal*, 2008, 44, 1023.
150. L. Shen, T.Y. Phong, L. Cheu, and I. Liu, *Polymer*, 2004, 45, 334.
151. D.Y. Yingwei, S. Iannace, I.E. Di Mazo, and L. Nicholas, *Journal of Polymer Science, Part B: Polymer Physics*, 2005, 6, 689.
152. H. Mirzazadeh, A.S. Katbal, and S. Bazgir, *Proceedings of the 8th International Conference Symposia on Polymer for Advanced Technologies*, Budapest, Hungary, September 13–16, 2005, paper 41, p. 2.
153. D. Yun Sheng, G. Hong Wei, and W. Seng Yang, *Polymer Materials Science and Engineering*, 2005, 21, 590.
154. R. Seldon, *Polymer Engineering and Science*, 1997, 37, 205.
155. Y.W. Leong, Z.A. Ishak, and A. Ariffen, *Journal of Applied Polymer Science*, 2004, 91, 3327.
156. H.S. Lee, P.D. Fasulo, W.R. Rodgers, and D.R. Paul, *Polymer*, 2005, 46, 11673.
157. Pascuel-Sanchez and V. Martinez, *Macromolecular Symposium*, 2006, 233, 137.
158. D.S. Choudhary, M.C. Jolland, and F. Sces, *Polymers and Polymer Composites*, 2004, 12, 383.
159. Y.H. Hu, E.-Y. Chen, and C.C. Wang, *Macromolecular*, 2004, 37, 2411.
160. D.W. Chae and B.C. Kim, *Journal of Materials Science*, 2007, 42, 1238.
161. S.U. Ryadiansyah, H. Ismail, and B. Azhari, *Polymer Composites*, 2008, 29, 1169.
162. C. Bartholom, E. Beyou, E. Bourgeat-Hamy, P. Cassagnau, P. Chanmonr, L. David, and N. Zydowicz, *Polymer*, 2005, 46, 9965.
163. O. Ruzimuradov, G. Rajan, and J. Mark, *Macromolecular Symposia*, 2006, 245, 322.
164. J.C.S. Norton, M.G. Han, P. Jiang, G.H. Shim, Y. Ying, S. Greager, and S.H. Foulger, *Chemistry of Materials*, 2006, 18, 4570.

165. T.P. Ngnyen, S.H. Yang, J. Gomes, and M.S. Wong, *Synthetic Metals*, 2005, 15, 269.
166. Y.W. Park and D.S. Lee, *Journal of Applied Polymer Science*, 2004, 94, 1780.
167. C.M. Lai, S.M. Sapuey Mahmid, M.A. Liuid, and Z.H.M. Dahlan, *Polymer Plastics Technology and Engineering*, 2005, 44, 619.
168. W. Leung, Z.A.M. Ishak, and A. Pirittin, *Journal of Applied Polymer Science*, 2004, 91, 3327.
169. M. Tasdemir and H. Yildorin, *Journal of Applied Polymer Science*, 2002, 83, 2967.
170. C.Y. Tang, L.C. Chan, J.Z. Liang, K.W.E. Cheng, and T. Wong, *Journal of Reinforced Plastics and Composites*, 2002, 21, 1337.
171. Z. Liang, *Journal of Applied Polymer Science*, 2007, 104, 1692.
172. X. Xu, Y. Sont, Q. Zheng, and G. Hu, *Journal of Applied Polymer Science*, 2007, 103, 2027.
173. B. Zhou, X. Ji, Y. Sheng, L. Wang, and Z. Jieng, *European Polymer Journal*, 2004, 40, 2357.
174. Q. Yan, W. Jiang, L. An, and R.K.Y. Li, *Polymer for Advanced Technologies*, 2004, 15, 409.
175. A.P. Gupta, U.K. Saroop, and M. Verma, *Polymer Plastics Technology and Engineering*, 2004, 43, 937.
176. F. Bianchi, A. Lazzeri, M. Pracella, A. Aguinu, and G. Ligeri, *Macromolecular Symposium*, 2004, 218, 191.
177. K. Wang, J. Wu, and H. Zeng, *Polymers and Polymer Composites*, 2006, 14, 473.
178. E.S.A. Rachid, K. Ariffin, H. Akit, and C.C. Kooi, *Journal of Reinforced Plastics and Composites*, 2008, 27, 1573.
179. Z.-H. Chen and K.-C. Gung, *Polymer Material Science and Engineering*, 2007, 23, 203.
180. B. Mary, C. Dubois, P.J. Garreau, and D. Brousgeay, *Rheologica Acta*, 2006, 45, 561.
181. J. Burghardt, N. Hansen, L. Hansen, and G. Hansen, *Proceedings of the SAMPE '06 Conference*, Long Beach, CA, 2006, paper 131.
182. A.S. Luyt, S.A. Molefi, and H. Krump, *Polymer Degradation and Stability*, 2006, 91, 1629.
183. J. Liu, B. Cankurtaran L. Weizorek, M.J. Ford, and M. Cortie, *Advanced Functions Materials*, 2006, 16, 1457.
184. N. Abdelaziz, *Journal of Applied Polymer Science*, 2004, 94, 2178.
185. D. Puryanti, S.H. Amed, A.M. Abdullah, and A.N.H. Yushaft, *International Journal of Polymeric Materials*, 2007, 56, 327.
186. J. Jancar, *Composite Interfaces*, 2006, 13, 853.
187. S. Mishra and N.G. Shimpi, *Journal of Applied Polymer Science*, 2007, 104, 2018.
188. M. Sambertsompop and A. Kositechi Yang, *Journal of Applied Polymer Science*, 2006, 102, 1896.
189. B.I. Shah, S.E. Selke, M.B. Walters, and P.A. Heiden, *Polymer Composites*, 2008, 29, 655.
190. E. Lezak, Z. Kulinski, R. Mansirek, A. Pirokowghe, M. Praeela, and K. Gadzinowikei, *Macromolecular Bioscience*, 2008, 8, 1190.
191. S.M. Luz, A.R. Goncalves, A.P. Del-Areo, and D.M. Ferrao, *Composite Interfaces*, 2008, 15, 841.
192. S. Joseph and S.J. Lumas, *Journal of Applied Polymer Science*, 2008, 109, 256.
193. S. Mahonty, S.K. Nayak, S.K. Verma, and S.S. Tripathy, *Journal of Reinforced Plastics and Composites*, 2004, 23, 625.
194. F. Lin, Y.C. Lu, J.L. Quo, and I.U. PAC, *Proceedings of the Polymer Symposium: Functional and High Performance Polymer Conference*, Taipei, November 14–16, 1994, p. 549.

195. G. Evrard, A. Belgrine, F. Corpier, E. Valot, and P. Dang, *Revenue Generale des Caeutichoucs et Plastiques*, May 1999, 777, 95.
196. Y.G. Jeong, S.C. Lee, and W.H. Jo, *Macromolecular Research*, 2006, 14, 416.
197. G.D. Sims, *International Standardisation of Coupon and Structural Element Test Methods*, ECCM Composites: Testing and Standardization 2, Hamburg, Germany, 1994.
198. ISO 3268, Plastics—GRP—Determination of tensile properties, 1978.
199. C. Bartholom, E. Beyou, E. Bourgeat-Hanny, P. Cassagnau, P.L. David, and N. Zydowicz, *Polymer*, 2005, 46, 9965.
200. C.-H. Chen and K.-C. Lein, *Polymers and Polymer Composites*, 2006, 14, 155.
201. M.R. Loos, S.H. Pezzin, S.C. Amico, C.P. Bergmann, *Journal of Material Science*, 2008, 18, 6064.
202. M.R. Loos, L.A.F. Coelho, S.H. Pezzin, and S.C. Amico, *Journal of Materials Research*, 2008, 11, 347.
203. B. Wong and G. Sun, *Polymer Engineering and Science*, 2007, 47, 1618.
204. S. Tzavalas, D.E. Mouzakes, and V. Drakonakis, *Journal of Polymer Science, Part B: Polymer Physics*, 2008, 46, 668.
205. M. Lai, J. Hi, J. Young, X. Tong, and H. Cheng, *Polymer International*, 2006, 53, 1479.
206. B. Wang, G. Sun, and J. Liu, *Polymer Engineering and Science*, 2007, 47, 1610.
207. S. Tzavalas, D.E. Mouzakis, V. Drakonakis, and V.E. Gregoriou. *Journal of Polymer Science, Part B: Polymer Physics*, 2008, 46, 668.
208. T. Aoki, Y. Ono, and T. Ogasawara, *Proceedings of the American Society for Composites, 21st Technical Conference*, Dearborn, MI, 2006, paper 31.
209. S.-M. Yuen, C.-C.M. Ma, C.-Y. Chunag, Y.-H. Hsiao, C.L. Chiang, and A.-D. Yu, *Composites, Part A: Applied Science of Manufacturers*, 2008, 39, 119.
210. S.-M. Yuen, C.-C.M. Ma, C.-C. Teng, H.-H. Wu, H.-C. Kuan, and C.-L. Chiang, *Journal of Polymer Science, Part B: Polymer Physics*, 2008, 46, 472.
211. ISO DIS 527—Part 5, Plastics—Determination of tensile properties—Test conditions for unidirectional fibre-reinforced plastic composites, 1994.
212. EN 61, Glass reinforced plastics—Determination of tensile properties, 1977.
213. prEN 2561, Carbon-thermosetting resin unidirectional laminates—Tensile test parallel to the fibre direction, 1989.
214. prEN 2747, Glass reinforced plastic—Tensile test parallel to the fibre direction, 1990.
215. ASTM D3039, Test method for tensile properties—Fibre-resin composites, 1993.
216. G.D. Sims, *Developments in Harmonisation of Standards for Polymer Matrix Composites*, EECM Composite Testing and Standardization, Amsterdam, Netherlands, 1992.
217. G.D. Sims, *Composites*, 1991, 22, 267.
218. BS 4994, Design and constructions of vessels and tanks in reinforced plastics, 1987.
219. BS 6464, Specifications for reinforced plastic pipes, fittings and joints for process plant, 1984.
220. ISO 10119, Carbon fibres—Determination of density, 1992.
221. ISO 10120, Carbon fibres—Determination of linear density, 1991.
222. ISO DIS 10548, Carbon fibres—Determination of size content and size amount, 2002.
223. ISO DIS 10617, Carbon fibres—Definition and vocabulary, 2010.
224. ISO 10352, Fibre reinforced plastics—Moulding compounds and prepregs, determination of mass per unit area, 1991.
225. PREN 2557, Carbon fibre preimpregnanates test methods for the determination of mass per unit area, Determination of mass unit area, 1998.
226. PREN 2339, Test methods for the determination of mass per unit area of woven textile glass—Fibre fabric preimpregnates, 1993.
227. P.T. Curtis, *Crag Test Methods for the Measurement of Engineering Properties of Fibre Reinforced Plastics*, 3rd ed., RAE TR 88012, 1988.

228. G.D. Sims, *Development of Standards for Advanced Polymer Matrix Composites: A BPF/ACS Overview*, NPL Report DMM (A), 1990.
229. G.D. Sims, *Standards for Polymer Matrix Composites, Part I: Assessment for Crag Test Methods Data*, NPL Report DMM (M)6, 1990.
230. G.D. Sims, *Standards for Polymer Matrix Composites, Part II: Assessment and Comparison of Crag Test Methods*, NPL Report DMM (M)7, 1990.
231. ISO 178, Plastics—Determination of flexural properties of rigid plastics, 1993.
232. ISO 8515, Textile glass reinforced plastics: Determination of comparison properties parallel to the laminate, 1991.
233. ISO 4585, Textile glass reinforced plastics—Determination of apparent interlaminar shear strength by short beam test, 1989.
234. ASTM D3518, In-plane shear stress-strain response of unidirectional reinforced plastics, 1991.
235. S. Kellas, J. Morton, and K.E. Jackson, *Damage and Failure Mechanisms in Scaled Angle-Ply Laminates*, ASTM STP 1156, 1993, p. 257.
236. prEN 6031, FRP—Determination of in-plane shear properties ± tensile test, 1995.
237. ISO 5725 (BSI 5497, 1987), Precision of test methods, Part 1: Guide for the determination of repeatability and reproducibility for a standard test method by inter-laboratory trial, 1986.
238. ISO CD 13586, Plastic—Determination of energy per unit area of crack (Gc) and the critical stress intensity factor (Kc), linear elastic fracture mechanics approach, 2000.
239. R. Khoshravan and C. Bathias, Delamination tests of DCB and ENF composite specimens and mode I and mode II, annual report, VAMAS WG5, 1990.
240. W.R. Broughton and A. Vamas, Round-robin project on polymer composite delamination in mode I and mode II; cyclic loading, NPL Report DMM(A)47, 1992.
241. G.D. Sims and A. Vanas, Round-robin on fatigue test methods for polymer matrix composites, Part I, Tensile and flexural tests of unidirectional material, NPL Report DMM(A)180, 1989.
242. G.D. Sims, Interim VAMS report on part 1 of polymer composites fatigue round-robin.
243. ASTM D3044, Standard test method for shear modulus of plywood, 1976.
244. S.W. Tasi and H.T. Hahn, *Introduction to Composite Materials*, Technomic Publishing Company, Lancaster, PA, 1980.
245. G.G. Sims, *International Standardisation of Coupon and Structural Element Test Methods*, ECCM Composites: Testing and Standardization 2, Hamburg, Germany, 1994.

4 Thermal Properties of Reinforced Plastics

This chapter reviews information available on the measurement of thermal properties of polymers, both unreinforced and reinforced. The consequences of incorporating reinforcing agents in the thermal properties of polymers are reviewed.

Information on standard methods for the determination of the properties of polymers is reviewed in Table 4.1. General reviews of the determination of thermal properties have been reported by several workers [1–6]. These include application of methods such as dynamic mechanical analysis [5], thermomechanical analysis [5], differential scanning calorimetry [4], thermogravimetric analysis [6], and Fourier transform infrared spectroscopy [4], in addition to those discussed below.

Thermal properties of a range of polymers are reviewed below, and some of these properties are compared in Table 4.2.

4.1 TYPICAL VALUES OF THERMAL PROPERTIES

Table 4.2 lists physical data, such as expansion coefficient, thermal conductivity, and specific heat, for a range of polymers. It is seen that expansion coefficients range from values as low as 2 mm/mm/°C × 10^{-5} (glass-filled epoxy resin), while for formaldehyde resins, phenol formaldehyde resin, polydiallyl phthalate polydiallyl isophthalate, and polyamide-imide this coefficient is as high as 20 mm/mm/°C × 10^{-5} (cross-linked polyethylene, perfluoroxy ethylene). Similarly specific heats range from as low as 1.0 kj/kg (polytrifluoroethylene) to as high as 2.3 kj/kg (low-density polyethylene).

In Table 4.3, critical temperature data related to polymers are quoted. These include heat distortion temperature, brittleness temperature, melt temperature, mold temperature, recommended maximum operating temperature, and performance at low temperature. Heat distortion temperatures, for example, range from 10–50 MPa (low-density polyethylene, polyurethane) to above 250 MPa (polyphenylene sulfide, urea formaldehyde-diallyl phthalate, polyether ether ketone, and polyamide-imide).

General discussions of the effect of reinforcing agents on the thermal properties of polymers include glass fiber–reinforced polyethylene terephthalate [28], multiwalled carbon nanotube–reinforced liquid crystalline polymer [29], polysesquioxane [30, 31], polyurethane [31], epoxy resins [32], polyethylene [33], montmorillonite clay–reinforced polypropylene [34], polyethylene [35], polylactic acid [36, 37], calcium carbonate–filled low-density polyethylene [38], and barium sulfate–filled polyethylene [39].

TABLE 4.1
Thermal Properties of Polymers

Method	ATS FAAR Apparatus Code Number	Units	Test Suitable for Meeting These Standards	Notes
Temperature of deflection under load or heat deflection or distribution temperature (Martens method)		°C	ASTM D648-04 [7]	
Softening point	10.01004	°C	ASTM D648-04 [7]	
VICAT	10.01009		ASTM D1525 [8]	
	10.01003		DIN EN ISO 306 [9]	
	10.01010		DIN EN ISO 75.1 [10]	
	10.01042		International Organization for Standardization (ISO) [14]	
			ISO 75-1 [10]	
			ISO 75-2 [12]	
			ISO 75-3 [12]	
			ISO 306 [13]	
			UNI EN ISO 75-1 [10]	
			UNI EN ISO 75-2 [11]	
			UNI EN ISO 75-3 [12]	
			UNI EN ISO 306 [13]	

Property	Value	Units	Standard	Notes
Melt flow index (or rate)	10.02013 10.02017	%	ASTM D1238 [15] ASTM D2116 [16] ASTM D3364 [17] UNI EN ISO 1133 [18]	
Low-temperature embrittlement (or brittle temperature)	10.12000 10.12006	°C	ASTM D746 [19] DIN ISO	
Melting temperature (Fisher-Johns)	7.0173	°C	ASTM	Also measured from different thermal analysis peaks
Mold shrinkage	Available	%	DIN 53464 [24–27]	
Thermal conductivity		w/m K	ASTM C177 [20] DIN 52612 [21]	
Linear expansion coefficient	10.4600	mm/mm cm/cm/°C in./in./°F	ASTM D696-44 [22] UNI 5284 [23]	
Specific heat		kj/kg	ASTM DIN	Can also be measured by differential scanning calorimetry
Minimum filling temperature	10.3500 10.35001 10.35002	°C	ASTM DIN	
Aging in air	10.69000		ASTM	

TABLE 4.2
Thermal Properties of Polymers

Polymer	Expansion Coefficient, mm/mm/°C × 10⁻⁵ ASTM D695 C177	Thermal Conductivity, °C 4.5 mm m² ASTM E177	Specific Heat, kj/kg	Mold Shrinkage, % (a) Injection (b) Compression	Softening Point—Vacat°C ASTM D635	Burning Rate, inc/min	Odor on Combination, mm/m/°C × 10⁻⁵
Low-density polyethylene	16–20	3.37	2.3	(a) 0.02–0.035	85–87	1.02–1.0	None
High-density polyethyelene	11–20	4.63–5.22	2.22–2.3	(a) 0.02–0.035	120–130	1.02–1.06	None
Polypropylene	6–11	1.38	1.93	(a) 0.015–0.015	150	0.75–0.83	None
Cross-linked polyethylene	20	—	—	—	—	—	—
Polymethyl-pentene	12	—	—	—	—	—	—
Polystyrene	7	—	1.34–1.45	(a) 0.002–006	82–103	—	None
Styrene–butadiene copolymer	3.4–23	0.42–1.26	1.34–1.45	(a) 0.002–0.008	78–100	—	Slight
Styrene–acrylonitrile	6.8	1.21	1.34–1.43	(a) 0.002–0.005	85–103	—	—
Polylonitrile-styrene-butadiene	7–10	—	—	—	—	—	—
Styrene-ethylene-butylene copolymer	16	—	—	—	—	—	—
Acrylate-styrene acrylonitrile copolymer		10	—	—	—	—	—
Fluorinated ethylene-propylene	56–23	2.1	2.8	(a) 0.001–0.001	288–295	—	None
Polytetrafluoroethylene	5.3–10.5	2.52	1.18	(a) — (b) 0.03–0.005	Transition point 372	—	None
Polytrifluorochloroethylene	9–22	—	1.05	—	Decomposition point 400	—	None
Polyvinyl fluoride	9	—	—	—	—	—	—
Polyvinylidene fluoride	16	—	—	—	—	—	—
Perfluoralkoxy ethylene	21	—	—	—	—	—	—
Ethylene tetrafluoroethylene copolymer	16	—	—	—	—	—	—

Rubber-modified polyvinyl chloride	5	1.89	0.84–2.1	(a) 0.005	78	—	None
Chlorinated polyvinyl chloride	7	—	—	—	—	—	—
Polyacetals (polyoxymethylene)	8.1–11	2.3	1.45	(a) 0.029	175	1–1.1	None
Polycarbonate	7–12	1.93	1.26	(a) 0.006–0.008	165	—	None
Ionomers	12	2.4	2.3	(a) 0.003–0.02	68–75	0.9–1.1	None
Polyphenylene oxide	5.2–6.0	3.2–6.0	1.23	(a) 0.005–0.007	124–148	—	None
Polyphenylene sulfide	—	—	—	—	—	—	—
Acrylics	—	1.68–5.52	—	(b) 0.001–0.004	80–98	0.6–1.3	None
Ethylene-vinyl acetate copolymer	5.9–18	0.3	1.47	(b) 0.001–0.004	64	—	None
Rigid epoxides	5.9	1.68–2.1	—	—	—	—	None
Flexible epoxides	5.9	1.68–2.1	—	—	—	—	None
Epoxy resins	0.05	—	—	—	—	—	—
Epoxy resins, glass filled	2	—	—	—	—	—	—
Phenol formaldehyde resins	2.5–6.0	1.26–2.52	1.6–1.76	(b) 0.01–0.02	—	—	None
Melamine formaldehyde resins	—	—	—	(b) 0.011–0.012	—	—	None
Urea formaldehyde resins	3	—	—	—	—	—	—
Polyurethane	15	—	—	—	—	—	—
Polybutylene terephthalate	12	—	—	—	—	—	—
Polyethylene terephthalate	8	—	—	—	—	—	—
Polydiallyl phthalate	3	—	—	—	—	—	—
Polydiallyl isophthalate	3	—	—	—	—	—	—
Polyarylates	6.3	—	—	—	—	—	—
Alkyd resins, mineral filler	3.75	—	—	—	—	—	—
Styrene-maleic anhydride copolymer	7	—	—	—	—	—	—
Polyether ether ketone	48	—	—	—	—	—	—
Polyimide	4.5	—	—	—	—	—	—
Polyetherimide	6.2	—	—	—	—	None	—
Polyesters, molding compound 1.1–5	9.4–10.6	1.04	—	—	—	—	—

Continued

TABLE 4.2 (*Continued*)
Thermal Properties of Polymers

Polymer	Expansion Coefficient, mm/mm/°C × 10⁻⁵ ASTM D695 C177	Thermal Conductivity, °C 4.5 mn m² ASTM E177	Specific Heat, kj/kg	Mold Shrinkage, % (a) Injection (b) Compression	Softening Point—Vacat°C ASTM D635	Burning Rate, inc/min	Odor on Combination, mm/m/°C × 10⁻⁵
Polyesters, sheet molding compound	2	—	—	—	—	—	—
Polyamide 6,6	8–10.15	2.18–2.43	1.65	(a) 0.015	264	—	None
Polyamide 6,10 (glass fiber reinforced)	3	—	—	—	—	—	—
Polyamide 6	8.13	—	1.68	(a) 0.009	215	—	None
Polyamide 11	9–15	4.2	2.4	(a) 0.012	185	—	—
Polyamide 12	10.4–11	2.9	2.1	(a) 0.003–0.015	175	—	None
Polyamide 6,12	9	—	—	—	—	—	—

TABLE 4.3
Critical Temperature Characteristics of Polymers

Polymer	Heat Distribution Temperature		Brittleness Temperature °C	Melt Temperature °C	Mold Temperature °C	Maximum Operating Temperature °C	Low Temperature Performance
	At 0.45 MPa °C	At 1.80 MPa °C					
Low-density polyethylene	35–50	35	Very good	220–260	20–40	50	Good
High-density polyethylene	45–82	37	Very good	220–260	20–60	50	Good
Polypropylene (20% C_aCo_3 filled)	90–105	68	Very poor	240–290	30–50	100	Fair
Cross-linked polyethylene	60	80	Good	150–170	20–80	90	—
Polymethylpentene	100	41	Very poor	260–320	40–70	75	—
Ethylene-propylene copolymer	93	54	Poor	310–350	100–150	60	—
Polystyrene	85	75	Poor	—	—	—	Poor-fair
Styrene-butadiene copolymer	67–93	73	Very poor	190–230	30–50	55	Poor-fair
Styrene-acrylonitrile copolymer	65–113	84	Very poor	220–270	50–80	55	Poor
Styrene-butadiene-acrylonitrile copolymer	88	82	Very good	250–270	70–100	70	—
Styrene-ethylene-butylene copolymer	20	20	Very good	200–240	50–70	85	—
Acrylate-styrene-acrylate-nitrile copolymer	97	85	Very poor	210–240	50–85	60	—
Fluorinated ethylene propylene	70	50	Very good	340–360	50–200	150	Excellent
Polytetrafluoroethylene	121	54	Excellent	—	—	180	Excellent
Polytrifluoroethylene	115	76	Good	270–300	90–110	130	—
Polyvinyl fluoride	121	82	Good	—	—	150	—
Polyvinylidene fluoride	121	82	Poor	280–320	80–90	130	—
Perfluoroxy ethylene	74	30	Good	74	30	170	—
Ethylene tetrafluoroethylene copolymer	105	71	Excellent	310–350	50–200	160	—
Rubber-modified polyvinyl chloride	71	—	—	—	—	—	Good

Continued

TABLE 4.3 (Continued)
Critical Temperature Characteristics of Polymers

Polymer	Heat Distribution Temperature		Brittleness Temperature °C	Melt Temperature °C	Mold Temperature °C	Maximum Operating Temperature °C	Low Temperature Performance
	At 0.45 MPa °C	At 1.80 MPa °C					
Polyacetal	160–169	110	Good	190–210	60–120	90	Good
Polycarbonate	138–150	60	Good	—	—	150	Excellent
Ionomers	38–43	—	Good	—	—	150	Excellent
Polyphenylene oxide	110–132	129	Poor	250–300	30–110	80	Good
Polyphenylene sulfide	217–260	—	—	—	—	—	—
Acrylics	71–95	—	—	—	—	—	Good
Ethylene vinyl acetate copolymer	20–64	20	Very good	160–220	20–40	50	Very good
Rigid epoxies	Up to 299	—	—	—	—	—	Good
Flexible epoxies	Up to 60	—	—	—	—	—	Very good
Epoxy resins	200	230	Very poor	—	—	130	—
Epoxy resins, glass filled	200	200	Very poor	60–80	160–190	130	—
Phenol formaldehyde	115–260	170	Very poor	60–80	150–170	150	—
Melamine formaldehyde	147						
Formaldehyde resin							
Polyurethane	20	20	Very good	195–230	20–55	70	—

Material							
Polybutylene terephthalate	150	60	Poor	230–270	30–90	120	—
Polyethylene terephthalate	72	70	Very poor	260–280	20–30	60	—
Polydiallyl phthalate	260	210	Good	60–90	150–180	160	—
Polydiallyl isophthalate	260	280	Good	60–90	150–180	180	—
Polyacrylate	171	170	—	—	—	130	—
Alkyd resins, mineral filled	260+	220	Poor	40–80	150–180	130	—
Styrene-maleic anhydride copolymer	110	105	Very poor	230–270	50–70	75	—
Polyether ether ketone	260	160	Very good	350–390	120–160	250	—
Polyimide	260	360	Good	—	—	360	—
Polyetherimide	210	200	Good	340–420	70–170	170	—
Polyesters, dough molding compound	205	—	—	—	—	—	Good
Polyesters, sheet molding compound	—	—	—	—	—	—	Good
Polyester resin	70	70–100	Good	—	—	—	—
Polyamide-imide	260	274	Poor	215–360	180–220	210	—
Polyamide 6,6	149–200	100	Very good	200	100	80	Good
Polyamide 6,10	149–220	220	Very good	250–280	60–100	—	Good
Polyamide 6	127–260	80	Very good	230–280	40–60	80	Good
Polyamide 11	144–182	55	Very good	200–280	40–60	70	Good
Polyamide 12	150	55	Good	190–210	40–60	70	Good
Polyamide 6,12	160	80	Very good	160	80	70	—

TABLE 4.4
Reinforced Polymers

Polymer	Reinforcing Agent	Application
Polypropylene	20% glass fiber	Cooling system expansion tank
Alkyd resin	Long glass fiber	Microwave ovens
Polyphenylene oxide	10% glass fiber	Components for heating systems
Polyether ether ketone	20% glass fiber	High-temperature applications
Polybutylene terephthalate	40% mineral and glass fibers	Heating appliances, ovens, grills
Polyamide 11	30% glass fiber	Heat-resistant components
Polyamide 12	30% glass fiber	Heat-resistant components
Polyetherimide	20% glass fiber	High-temperature connectors, heat exchanges
Polyetherimide	30% glass fiber reinforced	Thermal protectors
Polyphenylene sulfide	30% carbon fiber	High heat application
Polyphenylene sulfide	Glass fiber bead reinforced	High heat applications
Polyphenylene sulfide	40% glass fiber	High heat applications

4.2 HEAT-RESISTANT PLASTICS

Heat-resistant reinforced polymers are used in applications where good resistance to heat is important. Reinforced polymers meeting this requirement are listed in Table 4.4.

4.3 INDIVIDUAL THERMAL PROPERTIES

4.3.1 COEFFICIENT OF CUBICAL EXPANSION

Thermomechanical analysis is the measurement of dimensional changes (e.g., expansion, contraction, flexural, extension, volumetric expansion and contraction) in a material by movement of a probe in contact with the sample to determine temperature-related mechanical behavior in the range −180°C to 800°C as the sample is heated, cooled (temperature plot), or held at a constant temperature (time plot). It also measures linear or volumetric changes in the dimensions of a sample as a function of time and force.

The curve in Figure 4.1 illustrates a method of determination by this technique of the expansion coefficient of a polymer.

Table 4.2 shows that several polymers have a particularly high linear coefficient of expansion (i.e., >15 mm/°C × 10^{-5}), including polyolefins and some fluoropolymers. Not surprisingly, a high linear coefficient of expansion is associated with a high percentage of shrinkage in the mold. Conversely, polymers with a low coefficient of expansion have a low percentage shrinkage in the mold.

Subodh and coworkers [25] observed that the incorporation of ceramic powder filler into polytetrafluoroethylene composites had no effect on the polymer melting point, but as the filler level increased, the coefficient of thermal expansion decreased while the thermal conductivity increased.

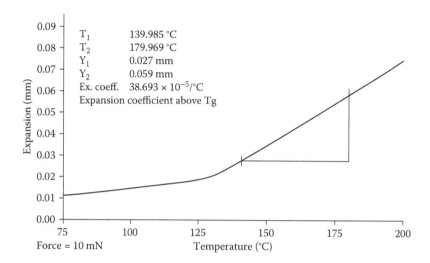

FIGURE 4.1 Thermomechanical analysis of epoxy printed board material. Measurement of expansion coefficient.

The addition of 20% glass fiber to an epoxy resin increased the coefficient of cubical expansion from 0.05 mm/mm 1°C × 10⁻⁵ for the base resin to 2 mm/mm/°C × 10⁻⁵ for the glass-filled resin.

Paul and Robeson [40] discuss problems associated with dimensional changes occurring during plastic molding operations. The latter is a particular concern for automotive parts where plastics must be integrated with metals that have much lower coefficients of thermal expansion. Fillers are frequently added to plastics to reduce the coefficient of thermal expansion. For low-aspect-ratio filter particles, the reduction in coefficient of thermal expansion follows, more or less, a simple additive rule and is not very large; in these cases, the linear coefficient of thermal expansion changes are similar in all three coordinate directions. However, when high-aspect-ratio fillers, like fibers or platelets, are added and well oriented, the effects can be much larger; in these cases, the coefficient of thermal expansion in the three coordinate directions may be very different.

The fibers or platelets typically have a higher modulus and a lower coefficient of thermal expansion than the matrix polymer. As the temperature of the composite changes, the matrix tries to extend or contract in its usual way; however, the fibers or platelets resist this change, creating opposing stresses in the two phases. When the filler-to-matrix modulus is large, the restraint to dimensional change can be quite significant within the direction of alignment. Platelets can provide their restraint in two directions, when appropriately oriented, while fibers can only do so in one direction. Because of their shape differences, fibers can cause a greater reduction in the direction of their orientation than platelets can. The coefficient of thermal expansion in the direction normal to the fibers or the platelet plane can actually increase when such fillers are added.

Montmorillonite platelets are particularly effective for reducing thermal expansion coefficients of plastics, as shown in Figure 4.2 for used exfoliated polyamide 6 nanocomposites.

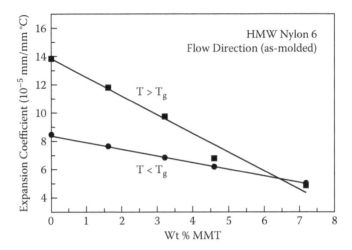

FIGURE 4.2 Coefficient of thermal expansion coefficient of montmorillonite-filled poly-amide nanocomposites.

These data measured the flow direction of injection-molded bars. When the semi-crystalline polyamide 6 matrix is above its glass transistor temperature, the reduction in the thermal expansion coefficient is greater than when it is below the glass transition temperature. Of course, the neat polyamide 6 has a higher coefficient of thermal expansion at temperatures above the glass transition temperature than when it is below the glass transition temperature, because of its lower modulus-to-glass transition temperature. The coefficients of thermal expansion of platelets are more effective for reducing the coefficient of thermal expansion.

Note that the two curves in Figure 4.2 seem to cross above the weight percent. For the specimens, the thermal expansion coefficient in the transverse direction is also reduced by the addition of montmorillonite clay, but not quite as efficiently as in the flow direction since the platelet orientation is not as great in the former case as it is in the latter case. The coefficient of thermal expansion in the normal direction actually increases as montmorillonite clay is added. These trends are quantitatively predicted by the theories discussed above [43].

Changes in the coefficient of thermal expansion occur when reinforcing agents are used in automotive applications. This is an obviously important consideration [41–44].

Sham and Kim [45] investigated the evolution of residual stresses during curing and thermal cycling of various epoxy resins by means of bimaterial strip bonding experiments within the chamber of dynamic mechanical test equipment. Resins tested were a neat epoxy resin, silica-modified underfill epoxy resin, and rubber-modified no-flow underfill epoxy resin. Changes in the viscosity and kinetics of curing of these resins with temperature were followed and related to residual stress evolution. The thermal expansion coefficients of the resins were also determined using a constitutive equation relating residual stress to changes in elastic modulus with temperature and compared with those obtained from thermomechanical analysis.

4.3.2 THERMAL CONDUCTIVITY

Details of the apparatus for conducting this measurement to ASTM C177 [48] and DIN 52612 [46] standards are given in Table 4.1.

Yu and coworkers [47] reported the results of thermal conductivity measurements on PS-aluminum nitride composites [47].

Dos Santos and Gregorio [48] measured the thermal conductivity of polyamide 6,6, polymethyl methacrylate, rigid polyvinyl chloride ether, and polyurethane foam.

Prociak [49] studied the effect of the method of sample preparation, the temperature gradient, and the average temperature of measurement on the thermal conductivity of rigid polyurethane foams blown with hydrocarbons and hydrofluorocarbons. The thermal insulation properties of different cellular plastics, such as rigid and flexible polyurethane foams, as well as expanded polystyrene, were compared. The thermal conductivity and thermal diffusivity of foams were correlated with the polyurethane matrix structure to demonstrate the effect of cell anisotropy on the thermal insulation properties of the rigid foams blown with cyclopentane and HFC-365/227 (93 wt% pentafluorobutane/7 wt% heptafluoropropane).

Various methods have been described for the determination of thermal conductivity. Capillarity has been used to measure the thermal conductivity of low-density polyethylene, high-density polyethylene, and polypropylene at various temperatures and pressure [50]. A transient plane source technique has been applied in a study of the dependence of the effective thermal conductivity and thermal diffusivity of polymer composites [51].

Transient hot-wire methods are most extensively used for the measurement of the thermal conductivity of polymers; these include polyamides, polymethyl methyl acrylates, polypropylene, polyvinyl chloride, low-density polyethylene, and polystyrene. This technique has been discussed by several workers [48, 52–57].

Other polymers for which thermal conductivity data are available include polyamide 6,6, polypropylene, polymethyl methacrylate, rigid polyvinyl chloride, cellular polyethylene [58], polyvinylidene fluoride [59], polyvinylidene difluoride-ceramic composite [60], polyethylene [61], and polyamide films [61].

Thermal conductivity studies have been conducted on a wide range of filled polymers and composites, including carbon fibers [62–68], aluminum powder [65], nitride [66], magnetite, barite, talc, copper, strontium ferrite [67], glass fiber–filled polypropylene and manganese or iron-filled polyaniline, carbon nanotubes [68], and nickel-cobalt-zinc ferrite in natural rubber [70].

Chiu and coworkers [69] measured the cylindrical orthotropic thermal conductivity of spiral woven fabric composites.

4.3.3 MELT TEMPERATURE AND CRYSTALLIZATION TEMPERATURE

Noroozi and Zelrajad [71] and others [72] conducted a detailed study of the effects of multiwalled carbon nanotubes on the thermal and mechanical properties of medium-density polyethylene matrix nanocomposites produced by mechanical milling.

Cold welding of medium-density polyethylene nanocomposites is much more extensive than that of neat medium-density polyethylene. This difference is attributed to the increased thermal conductivity of neat medium-density polyethylene when carbon nanotubes are added to the formulation.

4.3.4   SPECIFIC HEAT

Details are given in Table 4.1 for carrying out measurements to ASTM and DIN standards.

4.3.4.1   Hot-Wire Technique

This technique has been used to measure the specific heat of polyamide 6, polypropylene, polymethylene methacrylate, rigid polyvinyl chloride and cellular polyurethane foam, high-density polyethylene, low-density polyethylene and polystyrene [48], epoxy resins [73], polypropylene [74], polymethyl methacrylate [75], high-density polyethylene, low-density polyethylene, polypropylene and polystyrene [77], and polytetrafluoroethylene [76].

4.3.4.2   Transient Plane Source Technique

This technique has been used to measure specific heat fiber-reinforced phenol formaldehyde resins [78].

4.3.4.3   Laser Flash Technique

This technique has been used to determine the specific heat of polyvinylidene fluoride [59, 79].

Ranade et al. [80] showed that polyamide-imide-montmorillonite nanocomposites had a distinctly reduced specific heat when compared to the clay-free polymer.

4.3.5   MOLD SHRINKAGE

This indicates the inherent shrinkage of a material during injection molding. Molding conditions, orientation of fillers, orientation of flow, and other factors can influence the specific level of shrinkage, but the inherent quality is material dependent. This is the immediate level of shrinkage. Postmolding crystallization is accounted for by dimensional stability.

Apparatus for the measurement of this property according to DIN 53464 [24] is available from ATS FAAR (Table 4.1). Bertacchi and coworkers [26] reported computer-simulated mold shrinkage studies on talc-filled polypropylene, glass-reinforced polyamide, and a polycarbonate-acrylonitrile butadiene styrene blend.

4.3.6   THERMAL DIFFUSIVITY

Thermal diffusivity measurements have been reported for fiber-reinforced phenol formaldehyde resins, nylon 6,6, polypropylene, polymethyl methacrylate [81], and trifluoroethylene nanocomposites [82].

A laser flash technique has been used to determine the diffusivity of pyroelectric polymers such as polyvinylidene fluoride [83], whereas hot-wire techniques have been used to determine the thermal diffusivity of high-density polyethylene, low-density polyethylene propylene, and polystyrene [83]. Dos Santos and coworkers [84] utilized the laser flash technique to study the effect of recycling on the thermal properties of selected polymers. Thermal diffusivity expresses how fast heat propagates across a bulk material, and thermal conductivity determines the working temperature levels of a material. Hence, it is possible to assert that those properties are important if a polymer is used as an insulator, and also if it is used in applications in which heat transfer is desirable. Five sets of virgin and recycled commercial polymers widely used in many applications (including food wrapping) were selected for this study.

Disc-shaped samples (10 mm in diameter, thickness of 0.3–1 mm) were prepared by hot pressing the polymer or powder, or by cutting discs from long cylindrical bars. Measurements were carried out from room temperature up to ~50°C above the polymer crystalline T_m. Experimental results showed different behaviors for the thermal diffusivity of recycled polymers when compared with the corresponding virgin material.

4.3.7 Heat Distortion Temperature

4.3.7.1 Heat Distortion Temperature at 0.45 Mpa (°C)

The heat distortion temperature at 0.45 Mpa is the temperature that causes a beam loaded to 0.45 Mpa to deflect by 0.3 mm. If the heat distortion temperature is lower than ambient temperature, –20°C is given.

4.3.7.2 Heat Distortion Temperature at 1.80 Mpa (°C)

The heat distortion temperature at 1.80 Mpa is the temperature that causes a beam loaded to 1.80 to deflect by 0.3 mm. If the heat distortion temperature is lower than the ambient temperature, –20°C is given. Polymers such as low-density polyethylene, styrene ethylene-butene terpolymer, ethylene-vinyl acetate copolymer, polyurethane, and plasticized polyvinyl chloride distort at temperatures below <50°C, whereas others, such as epoxies, polyether ether ketone, polydiallylphthalate, polydiallyl isophthalate, polycarbonate, alkyd resins, phenol formaldehyde, polymide 6,10 polyimide, poly-etherimides, polyphenylene sulfide, polyethersulfone, polysulfonates, and silicones, have remarkably high distortion temperatures in the range of 150°C to >300°C. Thermomechanical analysis has been used to determine the deflection temperature of polymers and sample loading forces (i.e., plots of temperature vs. flexure).

Nam and coworkers [27] used dynamic mechanical analysis to measure the heat distortion temperature of polyphenylene sulfide/acrylonitrile-butadiene styrene.

Racz et al. [85], in their studies of polyimide-montmorillonite clay nanocomposites, measured heat distortion temperature, tensile and compact strength, and deformation.

The highest heat distortion temperature of 400°C flexural modulus (1.1×10^6 psi) and flexural strength (30,000 psi) of polyimide-montmorillonite clay nanocomposites made this polymer composite outstanding.

The tensile modulus and flexural modulus property values of phenol-resorcinol resin are reinforced with 35% glass fiber and are greatly superior to the properties of a 35% glass fiber–reinforced brominated resin bisphenol in bisphenol A-polyester resin.

Polyphenylene sulfide opens up new possibilities in aerospace, chemicals, electronic, food, and transportation applications [86]. Thermoformed polyphenylene sulfide, in addition to its excellent mechanical properties (Table 3.2), can tolerate high temperatures for prolonged periods of time. Thus, unreinforced polyphenylene sulfide has a thermal index of 186°C, and glass fiber–reinforced polyphenylene sulfide has a thermal index of 220°C.

4.4 MELTING TEMPERATURE (T_M) AND CRYSTALLIZATION TEMPERATURE

The melting temperature T_m is defined as the temperature at which crystalline regions appear in a polymer melt.

In semicrystalline polymers, some of the macromolecules are arranged in crystalline regions (termed crystallites), whereas the matrix is amorphous. The greater the concentration of crystallites (i.e., greater the crystallinity), the more rigid is the polymer (i.e., the higher is the T_m value).

Typical T_m values are reported in Table 4.5.

Figure 4.3 shows a sample temperature–heat flow differential scanning calorimetry (DSC) curve obtained for high-density polyethylene using a PerkinElmer DSC-7 instrument. It illustrates the measurement of the T_m and heat of melting in a single run.

The true T_m of crystalline polymers can be determined by plotting the DSC melting peak temperature as a function of the square root of the heating rate and making a linear extrapolation to the zero heating rate.

TABLE 4.5

Polymers with Particularly High Melting Temperature and Maximum Operating Temperatures

Polymer	Melting Temperature (T_m °C)	Maximum Operating Temperature (°C)
PEEK	350–390	250
Polyamide-imide	315–360	210
PEI	340–420	170
Ethylene-polytetrafluoroethylene	310–350	160
Fluorinated ethylene propylene	340–360	150
PPS	315–360	200
Polysulfones	310–390	150
PES	320–380	180

PEEK = polyether ether ketone, PEI = polyetherimide, PPS = polyparaphenylene sulfide, PES = x, PES = polyethersulphone.

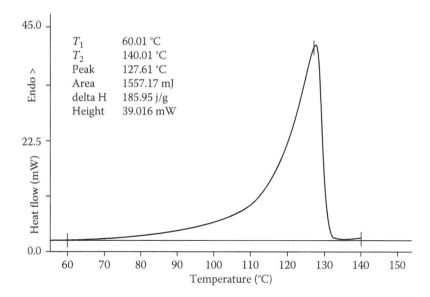

FIGURE 4.3 DSC measurement of T_g and heat of melting.

Differential thermal analysis has been used to study the effect of site chain length in polymers on T_m, as well as the effect of the heating rate of polymers on their T_m. DSC has been used to evaluate multiple peaks in polystyrene (PS) [86–88].

DSC measurements of the energy during melting of aqueous polymer solutions and gels yield heats of mixing, and sorption [89] has been used to study the heat changes occurring in a polymer as it is cooled (plots of temperatures vs. heat flow (measured in mW)).

McNally [90] used this technique to measure the melting temperature and crystallization temperature of medium-density polyethylene reinforced with various concentrations of milled multiwalled carbon nanotubes in the range of 0.5%–5%. The concentration of carbon in this range had hardly any effect on melting temperature or crystallization temperature; melting temperatures were 128.2°C for virgin polyethylene and 127.8°C polyethylene containing 5% carbon nanotubes. Similarly, crystallization temperatures were 113°C for neat polyethylene versus 115.2°C for polyethylene containing 5% carbon nanotubes.

Thermomechanical analysis has been used for softening measurements of polymers as well as the measurement of the amount of probe penetration into a polymer at a particular applied force as a function of temperature. This technique allows evaluation of T_m and the assessment of dimensional properties over the temperature range of use or under actual accelerated condition cycles (plots of temperature vs. compression (measure in mm)). The Fisher-Johns method has been applied for the determination of T_m.

Figure 4.4 demonstrates the application of differential scanning calorimetry to the determination of glass transition temperature.

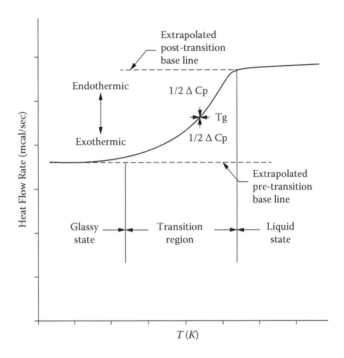

FIGURE 4.4 DSC run on high-density polyethylene (PerkinElmer DSC-7 instrument) showing measurement of T_m and heat of melting in a single run.

4.5 MAXIMUM OPERATING TEMPERATURE

Maximum operating temperatures are based on the Underwriters Laboratory (UL) rating for the long-term (100,000 h) ability for a material to sustain mechanical, electrical, and impact loads. Specifically, the UL temperature is defined as that temperature which causes the tensile strength of the material to fall to half its initial value after an exposure of 100,000 h.

In cases of short-term exposure (e.g., up to a few hours), which would include sterilization by autoclaving (typically 30 min at 134°C/270°C) or paint drying (20 min at 140°C/285°C), decisions made on the basis of maximum operating temperature may not be the most appropriate means of selecting candidate materials. For short exposure times, there may not be any significant level of oxidation or other chemical change in the material that would lead to a loss in mechanical or physical properties. However, even short exposure to high temperatures can lead to loss of dimensional stability. In some cases, therefore, in which the dimensional stability or stiffness of materials at the maximum use temperature is more important, it may be better to select materials on the basis of the heat distortion temperature.

Particularly high operating temperatures that can be sustained without impairment of physical properties are exhibited by particular plastics that have engineering applications. A high sustained melting temperature, T_m, is often a particular requirement of such plastics, as illustrated in Table 4.5.

TABLE 4.6

Assessment of Brittleness Temperatures

Very Good or Excellent	Poor
LDPE	PP
HDPE	Styrene-butadiene copolymer epoxies
Styrene-ethylene-butylene terpolymer	Acrylate-styrene-acrylonitrile
Ethylene-vinyl acetate copolymer	Chlorinated PVC
Polyamides	Plasticized PVC
PTFE	Unplasticized PVC
Ethylene-trifluoroethylene copolymer	Polymethylpentene
Fluorinated ethylene-propylene	—
Copolymer	—
Silicones	—

LDPE = low-density polyethylene, PP = polypropylene, HDPE = high-density polyethylene, PVC = polyvinyl chloride, PTFE = polytetrafluoroethylene.

Generally speaking, the inclusion of a glass fiber or carbon nanotube in the formulation will improve the thermal stability of polymers.

Melt flow studies have been reported on polypropylene [91], isotactic butadiene [92], glass fiber–filled polypropylene [93], and low-density polyethylene [94].

4.6 BRITTLENESS TEMPERATURE

The brittleness temperature is an indication of the subzero temperature at which the material becomes brittle. A *good* or *excellent* rating indicates a low brittleness temperature; that is, it is serviceable at low temperatures. A *poor* rating indicates a high brittleness temperature; that is, it has a limited subzero range of usefulness (Table 4.6).

An impact test for the measurement of brittleness temperature has been described by ASTM D746 [95].

4.7 MOLD TEMPERATURE

Seldon et al. [96] measured weld line impact strength and tensile strengths on injection-molded specimens of fiberglass–reinforced polyamide 6 and talc-reinforced polypropylene. The effect of hold pressure, impact velocity, melt temperature, and mold temperature on weld line strength was studied.

The effect of changing these parameters on the weld line strength was measured and compounded with strength via the weld line factor (defined as the strength of specimens within the weld line and those of specimens without a weld line). The highest weld line factor was obtained for unreinforced materials using a high melt temperature and holding pressure and a low mold temperature.

4.8 VICAT SOFTENING POINT

The VICAT method measures the temperature at which an arbitrary determination or a specified needle penetration (VICAT) occurs if polymers are subjected to an arbitrary standardized set of testing conditions (Table 4.1).

This property can also be measured by thermomechanical analysis (plot of temperature vs. compression (measured in mm)).

Tang and coworkers [97] found that the VICAT softening temperature of acrylonitrile-butadiene-styrene-calcium carbonate composites increased with the addition of the filler. This indicated the beneficial effect of the filler on the heat resistance of the terpolymer.

A ring ball apparatus has been used to determine a suitable temperature range, that is, softening temperature, for the extrusion of low-density polyethylene and high-density polyethylene polypropylene blends [98] and polyphenylene sulfide-acrylonitrile-butadiene blends.

4.9 GLASS TRANSITION TEMPERATURE

The glass transition temperature (T_g) is defined as the temperature at which a material loses its glasslike, more rigid properties and becomes rubbery and more flexible in nature. Practical definitions of T_g differ considerably between different methods; therefore, specification of T_g requires an indication of the method used.

Amorphous polymers, when heated above T_g, pass from the hard to the soft state. During this process, relaxation of any internal stress occurs. At the T_g, many physical properties change abruptly, including Young's and shear moduli, specific heat, coefficient of expansion, and dielectric constant. For hard polymeric materials, this temperature corresponds to the highest working temperature; for elastomers, it represents the lowest working temperature. Several methods exist for determining T_g. These include differential thermal analysis, differential scanning calorimetry, thermomechanical analysis, dilatometry, dynamic mechanical analysis, and nuclear magnetic resonance spectroscopic methods. Each method requires interpretation to determine T_g. For this reason, exact agreement is frequently not obtained between results obtained by different methods.

Reported T_g values of polymers cover a wide range, for example, −8°C to −100°C for polybutadienes and up to 100°C for polystyrene and polymethyl methacrylate.

4.9.1 DIFFERENTIAL SCANNING CALORIMETRY

Differential scanning calorimetry is probably the most frequently used method for determining T_g. In this method, a change in the expansion coefficient and the heat capacity occurs as a sample material is heated or cooled through this transition region.

Differential scanning calorimetry measures heat capacity directly, rapidly, and accurately, so it is an ideal technique for the determination of the glass transition temperature, T_g.

Figure 4.5 shows an idealized output obtained by differential scanning calorimetry. The T_g is taken as the midpoint in the thermogram as measured from the extension

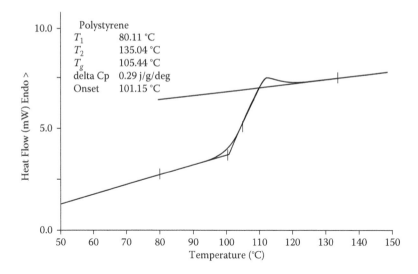

FIGURE 4.5 Determination of the T_g using the idealized output of differential scanning calorimetry.

of the pre- and posttransition baselines, that is, when the change in heat capacity assumes half the value of this change upon going through the transition. This choice is somewhat arbitrary. Other authors have suggested alternative techniques, such as taking the T_g at the first evidence of the displacement of the thermogram from the pre-transition baseline. The first of these methods is the most reliable and reproducible.

Alternatively, isothermal crystallization studies by differential scanning calorimetry provide a sensitive technique for measuring the molecular weight or structural differences between very similar materials (heat flow in mW) versus time plots. At the T_g, an endothermic shift occurs in the differential scanning calorimetry curve corresponding to the increase in specific heat (heat flow in mW vs. temperature). Determination of the T_g from the differential scanning calorimetry curve is based on the following three auxiliary lines: extrapolated baseline before the transition, inflection tangent through the greatest slope during transition, and baseline extrapolated after the transition. Figure 4.6 shows this measurement carried out on the polystyrene sample using a PerkinElmer DSC-7 instrument. It illustrates the temperature at which the differential scanning calorimetry has been applied to the determination of T_g various plastics, including polyimides [99], polyurethane [100], novolac resins [101], polyisoprene, polybutadienes, polychloroprene, nitrile rubber ethylene-propylene diene terepolymer, and butyl rubber [102], as well as bisphenol A epoxy diacrylate-trimethyl propane triacrylates [103].

4.9.2 THERMOMECHANICAL ANALYSIS

Plots of sample temperature versus dimensional (or volume) changes enable the T_g to be obtained. The T_g is obtained from measurement of sudden changes in the slope of the expansion curve.

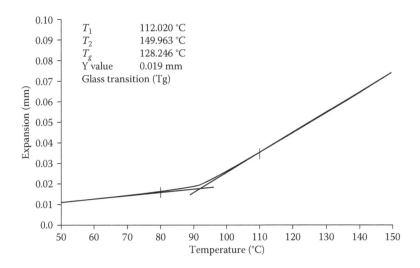

FIGURE 4.6 Determination of T_g of epoxy printed circuit board.

Johnston [104, 105] studied the effects of sequence distribution on the T_g of alkyl methacrylate-vinyl chloride and α-methylstyrene-acrylonitrile copolymers by differential scanning calorimetry, differential thermal analysis, and thermomechanical analysis.

Figure 4.6 shows an application of thermomechanical analysis to the characterization of a composite material, that is, an epoxy printed circuit board material. The T_g is readily determined from this curve.

When an elastomer was subjected to a penetration load of 0.03 N and a temperature range of −150 to 200°C, the material showed a slight expansion below the T_g, before allowing penetration at −17.85°C, resulting in a very marked T_g. The expansion that takes place at the T_g shows that the material is sturdy enough to resist further penetration even in its rubbery state.

Wohitjen and Dessy [106, 107] described a surface acoustic wave (SAW) device for conducting a thermomechanical analysis measurement of the T_g on polyethylene terephthalate, polytetrafluoroethylene bisphenol A, polysulfone, polycarbonate, and polymethyl methacrylate. There are several factors that distinguish the SAW device as a useful monitor of polymer T_g. The device is very sensitive, thereby permitting very small samples to be used. Preparation and mounting of the sample are simple and rapid. The device is quite rugged and possesses a small thermal mass, which permits fairly rapid temperature changes to be made.

Other methods that have been used to determine T_g include dynamic mechanical analysis [102, 103, 108–110], dielectric thermal analysis [102], nickel magnetic resonance spectrometry [111–116], and inverse gas chromatography [117, 118].

An increase in T_g accompanied by a reduction in thermal decomposition temperatures has been observed for the following: polyvinyl chlorate [119], polyvinyl pyridine [120], clay nanocomposites, polyamide 6, polyvinyl chloride [121], poly-4-v-vinyl pyridine [119], and fluoroelastomers.

TABLE 4.7
Glass Transition Changes with Nanofiller Incorporation

Polymer	Nanofiller	T_g (°C)
Polystyrene	SWCNT	3
Polycarbonate	SiC (0.5–1.5 wt%, 20–60 nm particles)	No change
Polyvinyl chloride	Exfoliated clay (MMT) (<10 wt%)	−1 to −3
Polydimethyl siloxane	Silica (2–3 nm)	10
Polypropylene carbonate	Nanoclay (4 wt%)	13
Polymethyl methacrylate	Nanoclay (2.5–15.1 wt%)	4–13
Polyimide	MWCNT (0.25–6.98 wt%)	−4 to 8
Polystyrene	Nanoclay (5 wt%)	6.7
Natural rubber	Nanoclay (5 wt%)	3
Poly(butylene terephthalate)	Mica (3 wt%)	6
Polyactide	Nanoclay (3 wt%)	−1 to −4

SWCNT = single-walled carbon nanotube, MMT = montmorillonite, MWCNT = multiwalled carbon nanotube.

Hu et al. [122] found that in silica-polymethyl methacrylate nanocomposites, the T_g increased with increasing silica content. Also, other thermal properties were enhanced by incorporating silica into the polymer matrix; for example, polymer degradation temperature was about 30°C higher than for virgin polymethyl methacrylate.

Zheng et al. [123] also showed that the T_g of silica-grafted acrylonitrile-butadiene-styrene terpolymers shifted markedly to high temperatures with increasing silica content, and that this was accompanied by improved thermal stability.

The addition of nanoparticles such as carbon nanotubes, clay, silica, and mica to polymers can cause an increase or decrease in the determined value of T_g. Table 4.7 illustrates positive or negative changes in T_g following the inclusion of various nanoparticles in the polymer formulation.

The glass transition of a polymer will be affected by its environment when the chain is within several nanometers of another phase. An extreme case of this is where the other environment is air (or vacuum). It has been well recognized in the literature that the T_g of a polymer at the air–polymer surface or thin films (<100 nm) may be lower than it is in bulk [124]. This can also be considered a confinement effect. A specific experimental example has been reported where poly(2-vinyl pyridine) showed an increase in T_g, polymethyl methacrylate showed a decrease in T_g, and polystyrene showed no change with silica nanosphere incorporation. These differences were ascribed to surface wetting [125]. The T_g decrease for polymethyl methylacrylate was ascribed to free volume existing at the polymer surface interface due to poor wetting. In most literature examples where T_g values have been obtained, usually only modest changes are reported (<10°C). In some cases, the organic modification of clay can result in a decrease in T_g due to plasticization [126]. It should be noted that the values noted in Table 4.7 involved relatively low levels of nanoparticle incorporation (<0.10 wt fraction and even lower volume fraction), and larger changes in T_g could be expected at much higher volume fraction loadings.

REFERENCES

1. M. Kato, A. Tsukigase, A. Usuki, T. Shimo, and U.T. Yazawa, *Journal of Applied Polymer Science*, 2006, 99, 3236.
2. H. Wa and J. Fun, *Polymer Testing*, 2008, 27, 122.
3. C. Maples and J. Parker, *Proceedings of the SAMPE 07, M&P Coast to Coast and around the World*, Baltimore, MD, 2007, paper 123, p. 7.
4. T. Tessuya, Y. Hashimoto, U.S. Ishiaku, M. Mozogucki, I.W. Leong, and H. Hamada, *Journal of Applied Polymer Science*, 2006, 99, 513.
5. N. Teramoto and M. Shibata, *Journal of Applied Polymer Science*, 2004, 91, 46.
6. A. Mohammed, U.L. Finkelstadt, P. Rayas-Duarte, D.E. Palmquist, and S.H. Gordon, *Journal of Applied Polymer Science*, 2009, 11, 114.
7. ASTM D648-04, Standard test methods for deflection temperature of plastics, under flexural load in the edgewise position, 2004.
8. ASTM 1525, Standard test method for VICAT softening temperature of plastics, 2000.
9. DIN EN ISO 306, Plastics—Thermoplastic materials—Determination of VICAT softening temperature, 2004.
10. DIN EN ISO 75-1, Plastics—Determination of temperature deflection under load—Part 1: General test methods, 2004.
11. DIN EN ISO 75-2, Plastics—Determination of temperature of deflection under load—Part 2: Plastics and ebonite, 2004.
12. ISO 75-3, Plastics—Determination of temperature of deflection under load—Part 3: High-strength thermosetting laminates and long-fibre-reinforced plastics, 2004.
13. ISO 306, Plastics—Thermoplastic materials—Determination of VICAT softening temperature (VST), 2004.
14. UNI 5642, Plastics—Thermoplastic materials—Determination of VICAT softening temperature, 2001. Superseded by UNI EN 306.
15. ASTM D1238-04c, Standard test method for melt flow rates of thermoplastics by extrusion plastometer, 2004.
16. ASTM D2116, Specification of FEP-fluorocarbon moulding and extrusion materials, 2003.
17. ASTM D3364-99, Test method for flow rates for poly(vinyl chloride) with molecular structural implications, 2004.
18. UNI EN 150 1133, Plastic determination of melt mass flat rate and the melt volume flow rate of thermoplastics, 2001.
19. ASTM D746, Test method for brittleness temperature of plastics and elastomers by impact, 2004.
20. ASTM C177, Standard test method for steady state heat flow measurements and thermal transition properties by means of the guarded hotplate apparatus, 2004.
21. DIN 52612, Testing of thermal insulating materials. Determination of thermal conductivity by means of the guarded hotplate apparatus, 1984.
22. ASTM D696-44, Test method for coefficient of linear thermal expansion of plastics between −30°C and 30°C with a vitreous silica dilatometer, 2003.
23. UNI 5284, Compassi per Scule: Recambi Portamina, 1994.
24. DIN 53464, Testing of plastics: Determination of shrinkage properties of mould materials from thermosetting moulding materials, 1962.
25. G. Subodh, M.V. Manjusha, J. Philip, and M.T. Sebastian, *Journal of Applied Polymer Science*, 2008, 108, 1716.

26. G. Bertacchi, A. Pipino, and G. Boero, *Macplas*, 2001, 26, 78.

27. J.-D. Nam, J. Kim, S. Lee, Y. Lee, and C. Park, *Journal of Applied Polymer Science*, 2003, 87, 661.

28. A. Pegoretti and A. Pencalti, *Polymer Degradation and Stability*, 2004, 86, 233.

29. S.K. Park, S.H., Kim and J.T. Hwang, *Proceedings of the 65th SPE Annual Conference*, Cincinnati, OH, May 6–11, 2007.

30. S.M. Yuen, C. Chi, M.A. Chih-Chim, S.E. Teng, H.-H. Wu, H.-C. Kuan, and C.-L. Chiang, *Journal of Polymer Science, Part B: Polymer Physics*, 2008, 46, 472.

31. W.P. Chang, S.-M. Yuen, H.-H. Wu, and T.-M. Lee, *Composites Science and Technology*, 2005, 65, 1703.

32. M.R. Loos, L.A.F. Pezzin, and S.C. Amico, *Materials Research*, 2008, 11, 347.

33. X.Q. Xiao, L.C. Zhang, and I. Zarudi, *Composites Science and Technology*, 2007, 67, 177.

34. W. Shaw and Q. Wang, *Polymer International*, 2005, 54, 336.

35. K. Stoeffler, P.G. Lafleur, and J. Denault, *Polymer Engineering and Science*, 2008, 48, 1449.

36. D.Y. Wei, S. Iannace, E.D. Matoa, and L. Nicolais, *Journal of Polymer Science, Part B: Polymer Physics*, 2005, 43, 689.

37. C.N. Shyang and L.G. Kuen, *Polymers and Polymer Composite*, 2008, 16, 263.

38. J. Zhao-Liang, *Journal of Applied Polymer Science*, 2007, 104, 1692.

39. F. Bianchi, A. Lazzeri, M. Pracella, A. D'Aguino, and G. Liperi, *Macromolecular Symposia*, 2004, 218, 191.

40. D.R. Paul and L.M. Robeson, *Polymer*, 2008, 49, 3187.

41. P.J. Yoon, T.D. Fornes, and D.R. Paul, *Polymer*, 2002, 43, 6727.

42. H.S. Lee, G.D. Fasulo, W.R. Rodgers, and D.R. Paul, *Polymer*, 2005, 46, 11673.

43. H.S. Lee, P.D. Fasulo, W.R. Rodgers, and D.R. Paul, *Polymer*, 2006, 47, 3528.

44. D.H. Kim, P.D. Fasulo, W.R. Rodgers, and D.R. Paul, *Polymer*, 2007, 48, 5308.

45. M.L. Sham and J.K. Kim, *Composites, Part A: Applied Science and Manufacturing*, 2004, 35A, No. 5, 537, 155N 1359-835X.

46. DIN 52612, Testing of thermal insulating materials. Determination of thermal conductivity of means of the guarded hotplate apparatus, 1984.

47. S. Yu, P. Hing, and X. Hu, *Composites, Part A: Applied Science and Manufacturing*, 2002, 33A, 289,

48. W.N. Dos Santos and R. Gregorio, *Journal of Applied Polymer Science*, 2002, 85, 1779.

49. A. Prociak, *Proceedings of a Rapra Technology Conference—Blowing Agents and Foaming Processes*, Shrewsbury, UK, 2004, p. 109.

50. A. Gothfleet and J. Sunder, *Proceedings of the 60th SPE Annual Technical Conference*, San Francisco, 2002, paper 236.

51. R. Agarwal, N.S. Saxena, K.B. Sharma, S. Thomas, and M.S. Speekala, *Journal of Applied Polymer Science*, 2003, 89, 1708.

52. G. Labudova and V. Vozarova, *Journal of Thermal Analysis and Calorimetry*, 2002, 67, 257.

53. Z. Zhang and M. Fuju, *Polymer Engineering and Science*, 2003, 43, 1755.

54. L. Halldahl, *Proceedings of the Rapra Technology Conference on Oilfield Engineering Polymers*, London, 2003, paper 22, p 259.

55. H. Kazeemine, *Journal of Applied Polymer Science*, 2003, 88, 2823.

56. H. Yan, N. Sada, and N. Toshima, *Journal of Thermal Analysis and Calorimetry*, 2002, 69, 881.

57. W.N. Dos Santos, *Polymer Testing*, 2005, 24, 932.

58. W.N. Dos Santos, R.G. Filho, P. Mummery, and A. Wallwork, *Science and Technology*, 2004, 14, 354.

59. W.N. Dos Santos, C.Y. Iguchi, and R. Gregorio, *Polymer Testing*, 2008, 27, 204.

60. G. Subod, M.J. Monjusham, J. Phillips, and M.T. Sabastian, *Journal of Applied Polymer Science*, 2008, 108, 1716.
61. W. Huijun and F. Jintu, *Polymer Testing*, 2008, 27, 122.
62. S.-Y. Fu and Y.-W. Mai, *Journal of Applied Polymer Science*, 2003, 88, 1497.
63. H.B. Shim, M.K. Seo, and S.J. Park, *Journal of Material Science*, 2002, 37, 1881.
64. R.D. Sweeting and Y.L. Lin, *Composites, Part A: Applied Science and Manufacturing*, 2004, 35A, 933.
65. O.M.A. Lupes and M.I. Felisberti, *Polymer Testing*, 2004, 23, 637.
66. S. Yu, P. Hing, and X. Hu, *Composites, Part A: Applied Science and Manufacturing*, 2002, 33A, 289.
67. Weidenfeller, M. Hofer, and F.R. Schilling, *Composites, Part A: Applied Science and Manufacturing*, 2004, 35A, 423.
68. G.P. Joshi, N.S. Saxena, T.P. Sharman, S. Verma, V. Dexit, and S.C.K. Mischre, *Journal of Applied Polymer Science*, 2003, 90, 430.
69. C.-H. Chiu, C. Chung-Li Huan, C.-C. Cheng, and K.H. Tsai, *Polymer and Polymer Composites*, 2007, 15, 167.
70. D. Puryant, S.H. Ahmed, M.H. Abdullah, and A.N.H. Yusoff, *International Journal of Polymeric Materials*, 2007, 56, 327.
71. M. Noroozi and S.M. Zebarjad, *Journal of Vinyl and Additive Technology*, 2010, 147.
72. M. Patricia, An investigation of the microstructure and properties of a cryogenically mechanically alloyed polycarbonate–polyethyl ether ketone system, PLS thesis, Virginia Polytechnic Institute and State University, 2001.
73. B.G. Chern, T.J. Moon, J.R. Howell, and W. Tan, *Journal of Composite Materials*, 2002, 36, 2001.
74. V.P. Privalku, V.B. Dolgoshey, E.G. Privalko, V.F. Shumsky, A. Lisougkii, M. Rodensky, and M.S. Eisen, *Journal of Macromolecular Science B*, 2002, B41, 539.
75. C. Chengzhi, K. Almdal, and J. Lyngaae-Jorgensen, *Journal of Applied Polymer Science*, 2004, 91, 609.
76. A.J. Papas and S. Cudzilo, *Journal of Thermal Analysis and Calorimetry*, 2004, 77, 329.
77. W.N. Dos Santos, R.G. Filtio, P. Mummery, and A. Wallwork, *Science and Technology*, 2004, 14, 354.
78. J.L. Haberfield and J.A. Reffner, *Society of Plastics Engineers*, 1975, 21, 858.
79. C.Y. Iguchi, W.N. Dos Santos, and R. Gregorio, *Polymer Testing*, 2007, 26, 788.
80. A. Ranade, N.H.D. D'Souza, and B. Grade, *Polymer*, 2002, 43, 3759.
81. *Rubber Asia*, 2003, 17, 93.
82. M. Aravind, C.W. Ong, and H.L. Chan, *Polymer Composites*, 2002, 23, 5, 925.
83. J.C.Y. Igushi, W.N. Dos Santos, and R. Gregorio, *Polymer Testing*, 2007, 26, 788.
84. W.N. Dos Santos, A. Marcondis, A. Agnelli, P. Mummery, and A. Wallwork, *Polymer Testing*, 2002, 26, 216.
85. L. Racz, B. Pukanszky, A. Pozsgay, and B. Puckanszky, *Progress in Colloid and Polymer Science*, 2004, 125, 96.
86. P.J. Lemstra, A.J. Schauken, and G. Challa, *Journal of Polymer Science, Part B: Polymer Physics*, 1972, 10, 230.
87. P.J. Lemstra, A.J. Schouter, and G. Challa, *Journal of Polymer Analysis: Physics*, 1974, 12, 1565.
88. A. Leary and C. Noel, *Journal de Chemie Physique-et-el Physico-Chemice Biologique*, 1972, 69, 875.
89. E. Ahad, *Journal of Applied Polymer Science*, 1974, 18, 1587.
90. T. McNally, *Journal of Polymers*, 2005, 46, 5222.
91. A. Godrillo, O.O. Santana, A.B. Martinez, M.L. Maspoch, M. Snaches-Soto, and J.I. Velasco, *Revista de Pla ticos Modernos*, 2002, 83, 395.

92. M. Le Bras, S. Bourbigot, C. Siat, and R. Delobel, in M. Le Bras, S. Bourbigot, G. Camino, and R. Delobel (eds.), *Fire Retardancy of Polymers*, Royal Society of Chemistry, Cambridge, UK, 1998, paper 54F, p. 266.
93. C.W. De Maio and S. Baushke, *Machine Design*, 2002, 74, 58.
94. N. Belhaneche-Bensemra and M.A. Chabou, *Proceedings of the ISFR Conference*, Ostend, Belgium, 2002, paper A01.
95. ASTM D746, Test method for brittleness temperature of plastics and elastomers by impact, 2004.
96. R. Seldon, *Polymer Engineering and Science*, 1997, 65, 7.
97. C.Y. Tang, L.C. Chon, J.Z. Liang, K.W.E. Cheng, and T.L. Wong, *Journal of Reinforced Plastics and Composites*, 2002, 21, 1337.
98. E.B.A.V. Pacheco and E.B. Mano, *Polimeros: Sciencia e Technologia*, 1996, 6, 22.
99. F.R. Diaz, J. Moreno, L.H. Tanlge, G.A. East, and D. Radic, *Synthetic Metals*, 1999, 100, 187.
100. R.G. Ferrliio and P.J. Achorn, *Journal of Applied Polymer Science*, 1997, 64, 191.
101. M.H. Alma and S.S. Kelly, *Polymer Degradation and Stability*, 2000, 68, 41.
102. A.K. Sircar, M.L. Glasaka, S. Rodriques, and R.P. Chartoff, *Proceedings of the 150th ACS Rubber Division Meeting*, Louisville, KY, 1996, paper 35, p. 65.
103. W.P. Yang, C. Wise, J. Wijaya, A. Gaeta, and G. Swei, *Proceedings of the RadTech '96 Conference*, Nashville, TN, 2006, vol. 2, p. 67.
104. N.W. Johnston, *Journal of Macromolecular Science—Chemistry*, 1973, A7, 53.
105. N.W. Johnston, *Macromolecules*, 1973, 6, 453.
106. H. Wohltjen and R. Dessy, *Analytical Chemistry*, 1979, 51, 1465.
107. H. Wohitjen and R. Dessy, *Analytical Chemistry*, 1979, 51, 1458.
108. R.G. Ferrino and P.J. Achorn, *Journal of Applied Polymer Science*, 1997, 64, 791.
109. M. Kluppel, R.H. Schuster, and J. Schaper, *Proceedings of the 151st Rubber Division*, Anaheim, CA, 1997, paper 54.
110. Z. Liang, P.K.Y. Li, and S.C. Tjong, *Journal of Thermoplastic Composite Materials*, 2000, 13, 13.
111. M.H. Alma and S.S. Kelley, *Polymer Degradation and Stability*, 2000, 68, 413.
112. L. Molu and D. Kisakurek, *Journal of Applied Polymer Science*, 2002, 86, 2232.
113. B. Wunderlich and D.M. Bodily, *Journal of Applied Polymer Science, Part C: Polymer Symposia*, 1964, 6, 137.
114. L. Reiger, *Polymer Testing*, 2001, 20, 199.
115. H.-Q. Zhang, W.-Q. Huang, C.-X. Li, and B.-L. He, *European Polymer Journal*, 1998, 34, 1521.
116. S.C. Tjong and W. Jiang, *Journal of Applied Polymer Science*, 1999, 73, 2247.
117. C.L. Beatty and Y.F. Froix, *Polymer Preprints*, 1975, 16, 628.
118. H.B. Smith, A.J. Manuel, and I.M. Ward, *Polymer*, 1975, 16, 57.
119. A. Renade, N.A. D'Souza, and B. Gnade, *Polymer*, 2002, 43, 3759.
120. C.Y. Lew, NR. Murphy, and G.M. McNally, *Proceedings of the 62nd Annual Conference (ANTEC 2004)*, Chicago, May 16–20, 2004.
121. P. Gong, M. Feng, C. Zhao, S. Zhang, and M. Yang, *Polymer Degradation and Stability*, 2000, 84, 289.
122. H. Hu, C.Y. Chen, and C.C. Wang, *Micromolecules*, 2004, 37, 2411.
123. K. Zheng, L. Chen, Y. Li, and P Lui, *Polymer Engineering and Science*, 2004, 44, 1077.
124. G. Guérin and R.E. Prud'homme, *Journal of Polymer Science*, 2007, 45, 10.
125. P. Rittigstein and J.M. Torkelson, *Journal of Polymer Science, Part B: Polymer Physics*, 2006, 44, 2935.
126. M. Pluta, J.K. Ieszka, and G. Boiteux, *European Polymer Journal*, 2007, 43, 2819.

5 Electrical Properties

5.1 TYPICAL ELECTRICAL PROPERTIES

The typical electrical properties of polymers are reviewed in Table 5.2.

Apparatus for carrying out these measurements are discussed in Table 5.1. The ATS FAAR teraohmmeter model M 150 E or equivalent is suitable for carrying out these measurements according to American Society for Testing of Materials (ASTM) D1401-02 [1], DIN IEC 60093 [2], and DIN IEC 60167 [3].

5.2 ELECTRICAL APPLICATIONS

Some of the electrical applications to which unreinforced and reinforced plastics have been applied are reported in Tables 5.3 and 5.4, respectively. It can be seen that applications cover a very wide area, ranging from wire and cable coverings, insulators, and transformers to their use in the fabrication of electronic and electrical components and the encapsulation thereof. The importance of electrical properties of polymers in relation to their applications cannot be overemphasized. Filled or reinforced polymers have distinct applications in certain critical applications.

5.3 INDIVIDUAL ELECTRICAL PROPERTIES

Various electrical properties of polymers are now discussed in further detail.

5.3.1 DIELECTRIC CONSTANT

The dielectric constant of a material is defined as the ratio of the capacitance of a particular capacitor containing the material to that of the same capacitor when the material is removed and replaced by air.

The dielectric varies with frequency and generally increases with temperature. The values quoted in Table 5.5 are typical room temperature values measured at a frequency of 1 kHz.

When designing a capacitor, it is desirable to include a material with a high dielectric constant because more power can be stored in a given volume. Conversely, materials with low dielectric constants are preferred for applications involving high frequencies to minimize electric losses.

Table 5.5 shows that some polymers have a particularly high dielectric constant; that is, these polymers have greater ability to store power in a given volume of polymer. These are polyesters, alkyd resins, phenol-formaldehyde, nylon 6,6, nylon 6,10, urea formaldehyde, and plastized polyvinyl chloride (PVC), all of which have dielectric constant values of >5 at 1 kHz. Tsuchiya and coworkers [14] discussed

TABLE 5.1

Apparatus Available from ATS FAAR for Determining the Electrical Properties of Polymers

Method	ATS FAAR	Measurement Units	Test Suitable for Meeting the Following Standards
Volume and surface resistivity	Teraohmeter	Ω cm (volume)	ASTM D257-99 [4]
	Model M510E	Ω (surface)	DIN IEC 60093 [2]
Dielectric constant and dissipation (power) factor	10.33510	Hz	ASTM D150-98 [5]
	10.33500		DIN 53483-1 [6]
Dielectric strength		kV UN	ASTM D149-97 [7]
Dielectric rigidity		V.mil	DIN 53481-1 [6]
Surface arc resistance		Seconds	ASTM D495-99 [8]
Tracking resistance	10.62000	Stage	DIN EN 60112 [9]
			CEI 15-50 [10]
			UNI 4290 [11]
Behavior during and after contact with a glowing rod		Stage	

the dielectric properties of polybinaphthalene ether, which demonstrates particularly good insulating properties.

Measurements of the dielectric constant have been reported on cross-linked polyethylene [15], low-density polyethylene [16], polyaniline nanofibers [17], and zirconium polyimide nanocomposites [18].

Insulating materials possess a dielectric constant (ε') characterizing the extent of electrical polarization that can be induced in the material by an electric field. If an alternating electric field is applied, the polarization lags behind the field by a phase angle, δ. This results in partial dissipation of stored energy. The dissipation energy is proportional to the dielectric loss (ε'), and the stored energy is proportional to the dielectric constant (ε'') permittivity.

The dielectric thermal analysis (DETA) technique normally obtains data from thermal scans at a constant impressed frequency. The T_g values at which molecular motions become faster than the impressed timescale are recorded as peaks in ε'' and tan δ.

It is a simple matter to multiplex frequencies over the whole frequency range 20–100 kHz, and under such conditions, the peaks in e^{11} are shifted to higher temperatures as frequency is increased. A further option allows data to be obtained in the frequency plane under isothermal conditions.

This technique is used principally for the rheological characterization of polymers and measurement of the dielectric constant.

The technique measures changes in the properties of a polymer as it is subjected to a periodic electric field. This produces quantitative data from which can be determined the capacitive and conductive nature of materials. Molecular relaxations can be characterized and flow and cure of resins monitored.

TABLE 5.2
Electrical Properties of Polymers

Polymer	Grade	Volume Resistivity (ohm/cm)	Dielectric Strength (mV/M)	Dielectric Constant (1 kHz)	Dissipation Factor (1 kHz) ASTM D150	Surface Arc Resistance (s)	Tracking Resistance DIN IEC 60112 [9]
LDPE	GP	16	27	2.3	0.0003	Very good	Excellent
Cross-linked polyethylene	GP	16	21	2.2	0.0004	Very good	Excellent
HDPE	GP	17	22	2.3	0.0005	Excellent	Excellent
PP	GP	15	18	2.6	0.0002	Excellent	Excellent
Polymethylpentene	GP	16	28	2.12	0.0002	Very good	Very good
Ethylene-propylene copolymer	GP	15	30	2.3	0.0005	Good	Very good
Styrene-butadiene copolymer	GP	15	12	2.5	0.0004	Poor	Very good
Styrene-ethylene-styrene copolymer	GP	16	26	2.2	0.006	Good	Good
PS	High impact	16	15	2.8	0.0006	Very poor	Very poor
Epoxy resins	GP	2	—	—	—	Poor	Poor
Acetals	GP	15	20	3.7	0.0015	Poor	Very good
Polyesters	GP bisphenol polyester	14	13	5.0	0.01	—	—
Polybutylene terephthalate	GP	15	20	3.2	0.002	Good	Good
PET	GP	16	40	3.5	0.01	Excellent	Poor
PEEK	GP	16.7	19	3.2	0.0016	Very good	Good
Diallyl isophthalate	GP	13	14	4.1	0.012	Excellent	Excellent
Diallyl phthalate	GP	14	16	4.4	0.01	Very good	Very good
Polyarylates	GP	16	15.1	3.13	0.005	—	—
Alkyd resins	GP	15	13	6.1	0.012	Very good	Very good
PC	GP	14	20	3.2	0.003	Very poor	Poor
PPO	GP	17	21	2.6	0.0004	Very poor	Poor

Continued

TABLE 5.2 (Continued)
Electrical Properties of Polymers

Polymer	Grade	Volume Resistivity (ohm/cm)	Dielectric Strength (mV/M)	Dielectric Constant (1 kHz)	Dissipation Factor (1 kHz) ASTM D150	Surface Arc Resistance (s)	Tracking Resistance DIN IEC 60112 [9]
Phenol-formaldehyde resins	GP	10	12	8	0.05	Very poor	Very poor
Styrene-maleic acid copolymer	GP	15	12	2.5	0.001	Poor	Poor
PMMA	GP	14	18	3.3	—	Excellent	Excellent
Ethylene-vinyl acetate	25% vinyl	16	27	2.9	0.02	Poor	Poor
Nylon-acrylonitrile alloy	GP	14	17	4.5	0.05	Very poor	Good
Polyamide 6,6	GP	15	25	8	0.2	Poor	—
Polyamide 6,10	Glass reinforced	15	17	3.8	0.015	Good	Poor
Polyamide 6	GP	14	25	8	0.2	Poor	Very good
Polyamide 6,9	GP	14	19	3.2	0.02	Poor	Poor
Polyamide 11	GP	14	20	4.0	0.05	Good	Very good
Polyamide 12	GP	15	60	3.6	0.05	Good	Poor
Polyamide 6,12	GP	14	20	3.6	0.02	Poor	Poor
Polyamide-imide	GP	17	23	3.5	0.01	Very poor	Very poor
PI	GP	16	22	3.6	0.0018	—	—
PEI	GP	15	24	3.15	0.0013	Very good	Poor
PU	GP	10	22	3.5	0.008	Good	Good
Urea-formaldehyde	GP	12	17	7	0.04	Good	Good

Material	Type						
Styrene acrylamide copolymer	GP	16	25	3	0.01	Good	Poor
Acrylonitrile-butadiene-styrene copolymer	GP	13	14	3.3	0.006	—	—
Acrylate-styrene-acrylonitrile copolymer	GP	14	22	3.2	0.025	Poor	Very poor
PTFE	GP	18	45	2.1	0.0001	Excellent	Excellent
Polyvinyl fluoride	GP	13	20	8	0.5	Very poor	Good
Polyvinylidene fluoride	20% carbon fiber reinforced	5	N/A	N/A	N/A	N/A	Very poor
Perfluoroalkoxy ethylene	GP	18	45	2.1	0.00012	Excellent	—
Ethylene-tetrafluoroethylene copolymer	GP	16	28	2.6	0.0008	Very poor	Very good
Ethylene-chlorotrifluoroethylene copolymer	GP	15	40	2.6	0.002	—	—
Fluorinated ethylene-propylene copolymer	GP	18	50	2.1	0.0002	Very good	Good
Chlorinated PVC	GP	14	14	3.1	0.025	Good	Poor
Ultra-high-molecular weight polyethylene	GP	18	28	2.3	0.0002	Very good	Excellent
Unplasticized PVC	GP	14	14	3.1	0.025	Very poor	Poor
Plasticized PVC	GP	13	30	5	0.06	Poor	Poor
PPS	Glass fiber	15.3	13.4	4.6	0.017	Very good	Poor
Polysulfone	GP	16	16.7	3.1	0.001	Good	Poor
PES	GP	17.5	16	3.5	0.0021	Very poor	Very poor
Silicones	GP	15	15.8	2.9	0.001	Very good	Excellent

LDPE = low-density polyethylene, HDPE = high-density polyethylene, PP = polypropylene, PET = polyethylene terephthalate, PEEK = polyether ether ketone, PC = polycarbonate, PPO = polyphenylene oxide, PMMA = polymethyl methacrylate, PI = polyimide, PEI = polyetherimide, PU = polyurethane, PTFE = polytetrafluoroethylene, PPS = polyparaphenylene sulfide, PES = polyethersulphone, GP = general purpose, N/A = not available.

TABLE 5.3

Electrical Applications of Unreinforced Polymers

Polymer	Application
Polypropylene	Electrical insulation
Epoxy resins	Electrical covering
Polyether ether ketone	Wire covering
Polydiallyl phthalate	Electrical connectors, switch gear housing, bush holders
Polycarbonate	Electrical switch gear
Polyethylene terephthalate	Electrical fittings, connectors
Polybutylene terephthalate	Electrical components, switches, motor components
Polyamide 6	Electrical connectors
Polyamide 11	Electrical mechanical components
Polyamide 12	Electrical and electronic applications, cable and wire covering
Polyimide	Capacitors, cable insulation, printed circuit boards
Polyetherimide	Electrical connectors
Polyamide-imide	Electrical connectors, printed circuit boards
Polyethersulfone	Electrical and electronic components
Polysulfone	Electrical and electronic components, switch housing
Perfluoroalkoxy ethylene	Wire coatings
Fluorinated ethylene-propylene	Electrical components

While the theory of dielectric analysis is well known, its use has long been frustrated by lack of effective instrumentation. Modern dielectric thermal analyses make the technique a practical reality.

Dielectric constant data have been reported on several unreinforced polymers, including polyimides [19–21], silicones [22], epoxy resins [23], polyurethanes [23], polyaniline nanofibers [16, 17], zirconium nanocomposites [24], cross-linked polyethylene [15], low-density polyethylene [22], doped polyimines [19], polyimide-3-zirconium propoxide nanocomposites [24], poly linu-naphthyl ether [30], and cross-linked polyethylene-polyethylene polyacrylate acid blends [15].

5.3.2 Dielectric Strength

Dielectric strength (MV/m) is the voltage that an insulating material can withstand before dielectric breakdown occurs. Dielectric strength generally increased with decreasing thickness of the specimens. The values given in Table 5.5 are typical dielectric strength values at room temperature for a 3-mm-thick specimen.

Details of the equipment needed for carrying out this measurement are in ASTM D140-97a [17].

The dielectric strength of various polymers under various categories covering the lower (10–20 Mv/m), intermediate (1–40 MV/m), and high (40–60 MV/m) ranges is shown in Table 5.5. Dielectric constants and arc and tracking resistances are also

TABLE 5.4

**Electrical Applications of Reinforced Plastics
(% glass fiber in parentheses unless otherwise stated)**

Polymer	Reinforcing Agent	Application
Polypropylene	Talc (20%)	Electrical system housings
Alkyl resins	Short glass fiber	Automotive, ignition, switches, relay bases
Epoxy resins	Glass fillers	Encapsulation of electronic and electrical device capitors, resistor protection, switch gear insulation
Polyether ketone	Glass fiber (20–35%)	Electrical components, printed circuit boards
Diallyl isophthalate	Long glass fiber	Switch gear brush holders
Diallyl phthalate	Long glass fiber	Electrical connectors, relays, switches
Diallyl phthalate	Short glass fiber	Connectors, relays, switch gear, brush holders
Polycarbonate	Short glass fiber (20–30%)	Electrical enclosures, relay separators
Polyethylene terephthalate	Glass fiber (30–40%)	Electrical components, terminal blocks
Polybutylene terephthalate	Glass fiber (10–20%)	Lamp sockets, switches, motor housings
Polyamide 6	Glass fiber (30%)	Electrical connectors, plugs
Polyamide 6,10	Glass fiber (30%)	Electrical plugs
Polyamide 7	Glass fiber (30%)	Electrical plugs, coils, and bobbins
Polyamide 8	Glass fiber (30–50%)	Electrical plugs, coils, and bobbins
Polyimides	Glass fiber (40%)	Terminal boards
Polyetherimide	Glass fiber (10–30%)	Electrical components, fuses, switches and printed circuit boards
Polyamide-imide	Glass fiber	Terminal strips, insulators
Polyphenylene sulfide	Glass fiber	Boxes for electronic circuits, relays, circuit breakers, terminal blocks, connectors, electric motor housings
Polyethersulfone	Glass fiber (30%)	Electrical components, printed circuit boards
Polysulfone	Glass fiber	Electrical components, printed circuit boards, electronic ignition components
Fluorinated ethylene-propylene	Glass fiber (20%)	Electrical components
Silicones	Glass fiber, mineral	Electronic component encapsulation
Polyamide 6,10	Carbon fiber (30%)	Electrical components
Polysulfone	Carbon fiber (30%)	Switching devices
Alkyd resins	Mineral and glass fiber	Switches, relay bases, capacitor encapsulation
Epoxy resins	Mineral and glass fiber	Encapsulation of electron and electrical devises, protection of capacitors and resistors
Silicones	Minerals and glass fiber	Electronic component encapsulation
Polydiallyl phthalate	Mineral	Relays, switches, communicators
Epoxy resins	Silica	Transformers, capacitor case sealing, electrical connectors

TABLE 5.5

Comparison of Dielectric Strengths, Dielectric Constant, and Resistance to Arcing and Tracking

Polymer	Dielectric Strength (mV/m)	Dielectric Constant (1 kHz)	Arcing Resistance	Tracking Resistance
Dielectric Strength, Range 10–20 mV/m				
PP	18	2.6	Excellent	Excellent
Styrene-butadiene	12	2.5	Poor	Good
PS	15	2.8	Very poor	Very poor
Polyesters	13	5.0	—	—
PEEK	19	3.2	Very good	Good
Polydiallyl isophthalate	14	4.1	Excellent	Excellent
Polydiallyl phthalate	16	4.4	Very good	Very good
Polyarylates	15	3.1	—	—
PC	20	3.2	Very poor	Poor
Alkyd resins	13	6.1	Very good	Good
Phenol-formaldehyde	12	8.0	Very poor	Poor
Styrene-maleic anhydride	12	2.5	Poor	Poor
PMMA	18	3.3	Excellent	Excellent
Nylon-acrylonitrile	17	4.5	Very poor	Good
Nylon 6,10	17	3.8	Good	Poor
Nylon 6,9	19	3.2	Poor	Poor
Urea-formaldehyde	17	7.0	Good	Good
Acrylonitrile-butadiene styrene	14	3.3	—	—
Chlorinated PVC	14	3.1	Good	Poor
Unplasticized PVC	14	3.1	Very poor	Poor
Phenylene sulfide	13	4.6	Very good	Poor
Polysulfone	16.1	3.1	Good	Poor
Polyethersulfone	16	3.5	Very poor	Poor
Silicones	5.8	2.9	Very good	Excellent
Dielectric Strength Range, 20–30 mV/m				
Ultra-high-molecular-weight LDPE	28	2.3	Very good	Very good
HDPE	22	2.3	Excellent	Excellent
Polymethylpentene	28	2.1	Very good	Good
Styrene-ethylene-styrene	26	2.5	Good	Good
Nylon 6,10	25	8	Poor	Very good
Polyamide-imide	22	3.5	Very poor	Very poor
PEI	24	3.1	Good	Poor
Dielectric Strength Range, 30–40 mV/m				
Ethylene propylene	30	2.3	Good	Very good
Plasticized PVC	30	5	Poor	Poor
Ethylene vinyl acetate	27	2.9	Poor	Poor
PET	40	3.5	Excellent	Poor

TABLE 5.5 (*Continued*)

Comparison of Dielectric Strengths, Dielectric Constant, and Resistance to Arcing and Tracking

Polymer	Dielectric Strength (mV/m)	Dielectric Constant (1 kHz)	Arcing Resistance	Tracking Resistance
Dielectric Strength Range, 40–60 mV/m				
Perfluoroalkoxy ethylene	45	2.1	Excellent	—
Ethylene trichlorofluoroethylene	40	2.6	—	—
PTFE	45	2.1	Excellent	Excellent
Fluorinated ethylene propylene	50	2.1	Very good	Good
Nylon 12	64	3.6	Good	—
Polyoxymethylene copolymer	69	3.7	Poor	Very good

TABLE 5.6

Applications of High-Dielectric-Strength Polymers

Polymer	Dielectric Strength (mV/m)	Arc Resistance	Tracking Resistance	Typical Application
Perfluoroalkoxy ethylene	45	Excellent	—	Insulation
Polyethylene terephthalate	40	Excellent	Poor	Electrical connectors
Ethylene trichlorofluoroethylene	40	—	—	Cable insulation
PTFE	45	Excellent	Excellent	High-temperature insulators, electronic relay
Fluorinated ethylene propylene	50	Very good	Good	Terminal blocking, wire insulation, electronic components
Polyamide 12	64	Good	—	Electrical components
Polyoxymethylene	69	Poor	Very good	Switch components

tabulated. In general, polymers that have a high dielectric strength (>40 MV/m) have excellent arcing and tracking resistance.

Polymers with the highest dielectric strength, between 30 and 70 mV/m, have been used in a range of critical applications, as listed in Table 5.6, which shows that these polymers also have excellent arcing properties and are used in critical electrical applications, such as connectors, high-temperature insulators, cable coverings, switch gear components, and electronic components.

Ramar and Alagar [28] compared the dielectric strengths of non-clay-reinforced ethylene-propylene-diene-tris(2-methoxyethoxy)vinyl silane–grafted and ethylene-propylene-polymethyl methacrylate blends. The values of dielectric strength, volume resistivity, surface resistivity, and arc resistance increased with increasing concentration of ethylene-propylene-diene-tris(2-methoxyethoxy)vinyl silane due to Si–O–Si linkages. The blends filled with nanoclay showed improved dielectric properties due to the presence of the inorganic moiety.

As an example, consider the selection of a plastic for use in the fabrication of an electrical component that meets the requirements of a minimum dielectric strength of 40 Mv/cm and minimum tensile strength 50 MPa. It is seen in Table 5.7 that polyethylene terephthalate, polyamide 12, and polyoxymethylene all meet this requirement.

Another requirement of a polymer for use in critical applications might be a combination of excellent continuous or maximum operating temperature (see Table 4.6) and good electrical properties. It is seen in Table 5.8 that polyether ketone, polyphenylene sulfide, and polytetrafluoroethylene all meet the requirements.

In Table 5.6, the polymers are grouped in ranges of increasing dielectric strength. The polymers with the highest dielectric strength, between 30 and 70 mV/m, have been used in a range of critical applications, as listed in Table 5.6. It is seen that these polymers, which also have excellent arcing properties, are used in the most critical electrical applications, such as connectors, high-temperature insulators, cable coverings, switch components, and electronic components. This is to be

TABLE 5.7
Selection of a Polymer with High Dielectric Strength and Tensile Strength

Polymer	Dielectric Strength (mV/m)	Tensile Strength (MPa)
Polydifluoro alkoxy ethylene	45	29
Polyethylene terephthalate	40	55
Polyethylene chlorofluoroethylene	40	30
Polytetrafluoroethylene	40	25
Fluorinated ethylene-propylene	50	14
Polyamide 12	64	50
Polyoxymethylene	69	50

TABLE 5.8
Comparison of Ranges of Electrical Properties and Continuous-Use Temperature

Use of Polyether Ketone Polyphenylene Sulfide and Polytetrafluoroethylene Temperatures

High continuous-use temperature	180–250°C
High dielectric strength	19–45 mV/mm
High dielectric constant	2.1–4.6 at 1 kHz
High arc resistance	Good to excellent
High tracking resistance	Good to excellent

contrasted with polymers having a low dielectric strength that are used in much less critical applications.

Cruz and Zanin [25] determined the dielectric strength of virgin and up to 10% recycled high-density polyethylene. The introduction of recycled polymer reduced dielectric strength by up to 17% due to metallic impurities in the recycled polymer.

5.3.3 VOLUME AND SURFACE RESISTIVITY

The volume resistivity (log, ohm, cm) of a material is its resistance to leakage of electrical current through itself. Specifically, it is the ratio of the potential gradient in the direction of the current to the current density. It is dependent upon moisture content and temperature. The values given in Table 5.2 are typical room temperature values.

Nedjar [26] have reported on the measurement volume of resistivity cross-linked polyethylene grades used in high-voltage cables and the effect of thermal aging on the electrical properties of the cable.

Volume resistivities are listed in Table 5.2. They range from as low as 2 ohm.cm for epoxy resins to as high as 16–18 ohm.cm for high-density polyethylene, polyether ether ketone, polystyrene, polymethylpentene, polyethylene terephthalate, polyarylates, polyphenylene oxide, polyamide imide, polyimides, polyurethane, polytetrafluoroethylene, perfluoroalkoxy ethylene, fluorinated ethylene-propylene copolymer, ultra-high-molecular-weight polyethylene, polysulfones, and polyethersulfones.

The volume resistivity, permittivity, and dielectric loss factor of nanostructured interpenetrating polymer networks based on natural rubber/polystyrene have been found to increase as a function of blend composition, reaching a maximum of 10^3–10^5 Hz dielectric loss factor [27]. Measurements of volume resistivity have also been reported on epoxy resin-polyaniline blends resulting in the establishment of a correlation between a shoulder on the 1583 cm^{-1} band with the degree of volume resistivity [31].

Niak and Mishra [13] found that the surface resistivity of natural fiber high-density polyethylene composites decreased with an increase of fiber content of the composite, whereas the volume resistivity increased.

5.3.4 ELECTRICAL RESISTANCE AND RESISTIVITY

Resistivity or specific resistance (S) is defined as the electrical resistance (R) multiplied by the cross-sectional area (A) of the polymer test piece divided by the length (L) (in cm):

$$G = RA/L$$

Electrical resistance measurements have been reported for polyaniline-cerium oxide composites [29], polyester fibers [30], and carbon-containing epoxy composites [31], whereas electrical positivity measurements have been reported for hydroelectron rheological polymers [32], reinforced acrylonitrile-butadiene-styrene [33], filled phenol-formaldehyde resin [34], and carbon black-polypropylene composites [35].

5.3.5 SURFACE ARC RESISTANCE

The surface arc resistance of a material is the time (s) taken for a conductive path to develop on the surface as the result of an arc passing between electrodes placed near the surface. Good surface arc resistance is necessary in conditions where high-voltage discharges are unavoidable. Surface arc resistance is reduced by fire retardants. Surface arc resistance is highly dependent upon the particular grade of material. This volume therefore ranks those materials that are inherently good or inherently poor. Therefore, the advice of the raw material supplier should be sought to determine the optimum grade of any material.

In Table 5.2,

- An *excellent* rating indicates excellent resistance to arcing.
- A *very poor* rating indicates poor resistance to arcing.
- *Not applicable* indicates that the material is conductive.

Details of the equipment for carrying out this measurement according to ASTM D495-99 [8] are given in Table 5.1.

Very few polymers have excellent surface arc assistance. These are high-density polyethylene polypropylene, polyethylene terephthalate, and polymethyl methacrylate.

5.3.6 TRACKING RESISTANCE

Tracking resistance (based on the KBr procedure) is the electrical potential applied between two electrodes placed in contact with the surface that causes a conductive path to develop between them. Good tracking resistance is necessary in applications in which contamination and degradation of the surface are unavoidable, for example, in marine environments. Tracking resistance is highly dependent upon the particular grade of material. The advice of the materials supplier should be sought to determine the optimum grade of any material. The presence of fire retardants can severely reduce the tracking resistance of a material:

- An *excellent* rating indicates excellent resistance to tracking (e.g., polyolefins).
- A *very poor* rating indicates poor resistance to tracking.
- *Not applicable* indicates that the material is conductive (e.g., epoxies, polystyrene).

The apparatus supplied by ATS FAAR (Table 5.1) evaluates, during a short period of time, the low-voltage (≤600 V) track resistance (or comparative tracking index) of insulating materials.

The tracking resistance test is conducted by applying two electrodes—with a force of 1 N—to the surface of the specimen. A 0.01% solution of ammonium chloride is dropped at the rate of one drop, having a volume of 20 ± 2 mm^3, every 30 s, while a variable-tension 50 Hz AC voltage is applied. The test ends when the sort circuit current is 1 A.

Kevlar para-aramide fibers have been shown to have three times the tracking resistance of general-purpose rubber [36]. Sechin et al. [37] carried out tracking resistance measurements on polyaniline-cerium oxide composites.

5.3.7 DISSIPATION FACTOR

The dissipation factor (or dielectric loss) is a measure of the fraction of energy absorbed per cycle by insulation from the field at a frequency of 1 kHz or 50 Hz. A low dissipation factor implies efficient electrical insulation.

The dissipation factor is defined as the ratio of in-phase power. It may also be defined as the tangent of the loss angle. It is dependent upon frequency. The lower the dissipation factor, the more efficient is the insulator system:

- An *excellent* rating indicates a low dissipation factor.
- A *very poor* rating indicates a high dissipation factor.
- *Not applicable* indicates that the material is conductive.

Dissipation factors (5.2) range from very low, for example, polyolefins (i.e., excellent rating) to relatively high, for example, nylons and polyvinyl fluoride. Methods for the measurement of the dissipation factor are listed in Table 5.1.

5.3.8 ELECTRICAL CONDUCTIVITY

Some measurements of this property have been made in a range of electrically conducting polymers. These include epoxy resin/polyaniline-dodecylbenzene sulfonic acid blends [38], polystyrene-black polyphenylene oxide copolymers [38], semiconductor-based polypyrroles [33], titanocene polyesters [40], boron-containing polyvinyl alcohol [41], copper-filled epoxy resin [42], polyethylidene dioxy thiophene-polystyrene sulfonate, polyvinyl chloride, polyethylene oxide [43], polycarbonate/acrylonitrile-butadiene-styrene composites [44], polyethylene oxide complexes with sodium lanthanum tetra-fluoride [45], chlorine-substituted polyaniline [46], polyvinyl pyrolidine-polyvinyl alcohol coupled with potassium bromate tetrafluoromethane sulfonamide [47], doped polystyrene block polyethylene [38, 39], polypyrrole [48], polyaniline-polyamide composites [49], and polydimethyl siloxane-polypyrrole composites [50].

5.4 EFFECT OF GLASS FIBER REINFORCEMENT ON ELECTRICAL PROPERTIES OF POLYMERS

The incorporation of 10%–60% of reinforcing agents such as glass or carbon fiber into polymers can have a profound effect on their electrical and mechanical properties.

In general, 10%–60% glass fiber causes improvement in the dissipation factor in polycarbonate and polyamide 6,6 and in the surface arc resistance in polyphenylene oxide and epoxy resins, but causes deterioration of the dielectric constant in poly-amide 6,6 and in the tracking resistance of polypropylene epoxies and phenylene oxide. Glass fiber caused an improvement in the dielectric strength of epoxies.

As shown below, polyether ether ketone has excellent electrical properties, rendering it suitable for critical applications such as electrical plugs and electrical gear housings and printed circuit board manufacturers.

Volume resistivity: 16.7 ohm.cm
Dielectric strength: 19 mV/m
Dielectric constant at 1 kHz: 3.2
Surface arcing resistance: Very good
Tracking resistance: Good
Dissipation factor: 0.0016

For some applications, however, an improvement in the mechanical properties of polyether ether ketone would be desirable. It has been found that the incorporation of 30% glass fiber into the formulation produces a distinct improvement in mechanical properties without compromising electrical properties:

	Unreinforced	30% Glass Fiber Reinforced
Tensile strength, MPa	92	151
Flexural strength, GPa	3.7	10.3
Elongation at break, %	50	2.2
Strain at yield, %	4.3	—
Izod impact strength, 1 kj/m	0.083	0.09

Li et al. [51] investigated the effect on volume resistivity brought about by the incorporation of glass fibers into polypropylene-epoxy resin composites.

In Table 5.9 the mechanical, thermal, and electrical properties of virgin and 20% glass fiber–reinforced perfluoroalkoxy ethylene are compared.

Volume resistivity and dielectric strengths are virtually unaffected by the incorporation of glass fiber, while the dielectric constant undergoes a moderate increase.

Additionally, glass fiber reinforcement produces a significant increase in heat distortion temperature and a moderate increase in tensile strength but, as might be expected, a severe drop in percentage elongation at break.

5.5 EFFECT OF CARBON FIBER REINFORCEMENT ON ELECTRICAL PROPERTIES OF POLYMERS

Only a limited amount of quantitative data are available on the effect of carbon fiber on the electrical properties of reinforced polymers, as opposed to unreinforced plastics, where data are readily available.

TABLE 5.9

Properties of Unreinforced and 20% Glass Fiber–Reinforced Perfluoroalkoxy Ethylene

	Unreinforced	20% Glass Fiber Reinforced
Maximum operating temperature, °C	170	170
Water absorption, %	0.03	0.04
Tensile strength, MPa	29	33
Flexural modulus, GPa	0.7	0.7
Elongation at break, %	300	4
Strain at yield, %	85	—
Notched Izod, kj/m	1.06+	0.7
Heat distortion temperatures at 0.45 MPa°C	74	110
Heat distortion temperatures at 1.80 MPa°C	30	58
Material mold shrinkage, %	4	—
Surface hardness	SD60	SD60
Linear expansion, m/m/°C × 10^{-5}	21	—
Flammability, UL94	VO	—
Oxygen index, %	95	—
Volume resistivity log, ohm/cm	18	—
Dielectric strength, mV/M	45	—
Dielectric constant, 1 kHz	2.1	—
Dissipation fact, 1 kHz	0.0002	—
Melt temperature, °C	360–420	500–550
Mold temperature range, °C	50–250	100–300

In general, carbon fibers cause a deterioration of volume resistivity in epoxy resins and polytetrafluoroethylene, improve the dielectric strength and dielectric constant in epoxy resins, and cause deterioration in the surface arc resistance and tracking resistance.

Carbon fibers cause a deterioration of volume resistivity, and arc resistance in polyamide 6,6. Mineral fillers also decrease the dielectric constant and surface arc resistance in polyamide 6,6. This is confirmed by a comparison of electrical, mechanical, and thermal properties of polyoxymethylene, where the incorporation of 30% carbon fiber reduces volume resistivity and dielectric strength, in addition to a distinct improvement in tensile strength at heat distortion temperature (Table 5.10).

Nishikawa and coworkers [52] and Gao and Zhao [32] investigated the resistivity of acrylonitrile-butadiene-styrene terpolymer reinforced with crushed carbon fiber and elastic gels based on gelatin/starch, respectively.

Li and coworkers [51] investigated the effect of the glass fiber and epoxy content of carbon black–filled polypropylene(epoxy) glass fiber composites on volume resistivity.

Drubeski et al. [35] investigated the electrical properties of electrically conductive hybrid carbon fiber–carbon black fiber polypropylene composites. These properties were evaluated by resistance measurements and scanning electron microscopy, respectively.

TABLE 5.10

Effect of Carbon Fiber Addition on Electrical, Mechanical, and Thermal Properties of Polyoxymethylene

	Virgin Polymer	30% Carbon Fiber–Reinforced Polymer
Tensile strength, MPa	50	85
Flexural modulus, GPa	27	17.2
Elongation at breach, %	20	1
Strain at yield, %	8	N/A
Heat distortion temperature at 0.45 MPa°C	160–169	188
Heat distortion temperature at 1.8 MPa°C	110	162
Mold shrinkage, %	0.025	0.02
Expansion coefficient, mmg/mm/°C × 10^{-5}	8.1–11	4
Maximum operating range, °C	90	170–210
Mold temperature	60–120	100–120
Volume resistivity, ohm/cm	15	2.95
Dielectric strength, mV/m	20	3.3
Dielectric constant, 1 kHz	3.7	N/A
Dissipation factor, 1 kHz	0.0015	N/A

N/A = not applicable.

Synergism between the conductive networks was examined, and the effects of fiber orientation and fiber length on composite resistance were evaluated. The resistance of the composites containing both carbon fiber and carbon black fillers was also compared with that of composites containing only carbon black or carbon fibers.

Electrical properties have been reported on numerous carbon fiber–reinforced polymers, including carbon nanofiber–modified thermotropic liquid crystalline polymers [53], low-density polyethylene [54], ethylene vinyl acetate [55], wire coating varnishes [56], polydimethyl siloxane polypyrrole composites [50], polyacrylonitrile [59], polycarbonate [58], polyacrylonitrile-polycarbonate composites [58], modified chrome polymers [59], lithium trifluoromethane sulfonamide–doped polystyrene-block copolymer [60], boron-containing polyvinyl alcohols [71], lanthanum tetrafluoride complexed ethylene oxide [151, 72, 73], polycarbonate-acrylonitrile diene [44], polyethylene deoxythiophenel, blends of polystyrene sulfonate, polyvinyl chloride and polyethylene oxide [43], polypyrrole [61], polypyrrole-polypropylene-montmorillonite composites [62], polydimethyl siloxane-polypyrrole composites [63], polyaniline [46], epoxy resin-polyaniline dodecyl benzene sulfonic acid blends [64], and polyaniline-polyamide 6 composites [49].

5.6 EFFECT OF CARBON BLACK REINFORCEMENT ON ELECTRICAL PROPERTIES OF POLYMERS

Tantaway et al. [65] investigated the effect of Joule heating on the electrical and thermal properties of conductive carbon black epoxy resin composites.

The Joule heating effect was shown to be an effective and promising method for enhancing the electrical and thermal stabilities of epoxy resin-carbon black composites for consumer use as heaters and in other electronic areas, such as electromagnetic shielding.

Jurkawska et al. [66] investigated the effect of fullerene and carbon black on the properties of rubber. In addition to beneficial improvements to physical properties such as elastic modulus fullerene at the 0.06–0.75 phr level, also affected were electrical properties such as dielectric loss angle and permittivity.

Zhou et al. [67] investigated the effect on electrical properties of incorporating carbon black in a low-density polyethylene composite and low-density polyethylene ethylene methyl acrylate blends. Electrical conductivity/resistivity measurements have shown that the percolation threshold of ethylene-methylacrylate blend polymer composites was significantly lower than that of the low-density polyethylene composite, although in an ethylene-methyl acrylate composite the threshold is higher. The effect was due to preferential absorption of the carbon black into low-density polyethylene due to phase separation and immiscibility in low-density polyethylene-ethylene-methyl acrylate blends. Viscosity of polymers in the blend appeared to determine distribution on the carbon black, indicating that choice of polymer viscosity could be used to control carbon black distribution.

Yin et al. [68] investigated the critical resistivity, dispersivity, and percolation threshold of low-density polyethylene carbon black. Li et al. [69] investigated the electrical properties and crystallization behavior of four different kinds of carbon black–filled polypropylene composites, prepared by the melt mixing method. All showed typical characteristics of percolation, but noticeably different percolation thresholds. When using carbon black with a higher structure, smaller particle diameter, and larger surface area, the composite showed better electrical conductivity and a lower percolation threshold.

A study of crystallization behavior by differential scanning calorimetry showed that carbon black particles in polypropylene were able to act as nucleating agents that accelerated the crystallization rate. Isothermal crystallization kinetics studies indicated that the crystallization kinetic constant and the rate of crystallization of the polypropylene-carbon black composites were higher than those of polypropylene. The value of the Avrami exponent indicated three-dimensional spherulitic growth in a spherical form. The free energy of chain folding for polypropylene crystallization of the carbon black–reinforced polypropylene composites was lower than that of polypropylene itself, and decreased further with increasing carbon black content.

Pourabas and Peyghambardoost [70] prepared positive temperature coefficient composites by using metal-modified and unmodified carbon black in a matrix of high-density polyethylene. Modification with metallic particles led to properties that were related to changing the surface properties of the carbon black. The intrinsic electrical conduction of carbon black also changed after modification. These changes in properties endowed some desirable characteristics to the positive temperature coefficients.

Chen et al. [71] investigated the electrical and mechanical properties of carbon black-polyvinylidene fluoride-tetrafluoroethylene-propylene films cross-linked with triethylene tetramine. These are possible contenders as a binder for lithium-ion battery electrodes.

The mechanical properties of poly(vinylidene fluoride-tetrafluoroethylene-propylene) were studied as a function of the degree of cross-linking. Triethylenetetramine and bisphenol (the latter as incorporated into the polymer by the manufacturer) were used as cross-linkers to improve the mechanical properties of polyvinylidene fluoride-tetrafluoroethylene propylene and carbon black–filled polymer. Cross-linking was carried out by adding magnesium oxide and calcium oxide, followed by heating. Bisphenol cross-linking improved the mechanical properties, but further cross-linking using triethylene tetramine was needed to obtain good extensibility under repeated deformation of carbon black polyvinylidene fluoride-tetrafluoroethylene polypropylene composites. Adding cross-links had little or no effect on the electrical properties of the composite films. The cross-linked polymer did not become brittle in solvent, which is typically used in lithium battery electrolytes, so this binder system may be used for electrode materials that show large-volume changes during charge–discharge cycling.

Electrical resistance measurements have also been carried out on composites of carbon black with polypropylene [35], composite fibers, and epoxy resins [91].

5.7   EFFECT OF GRAPHITE REINFORCEMENT ON ELECTRICAL PROPERTIES OF POLYMERS

Beloshenko et al. [31] measured the electrical resistance of epoxy composites containing either thermoexpanded graphite alone or thermoexpanded graphite and kaolin under uniaxial compression and over the shape recovery temperature range. The data obtained revealed that shape recovery was accompanied by a jumplike increase in electrical resistance, which was related to a change in sample length, and that the addition of kaolin gave rise to a decrease in electrical resistance and an increase in the value of the electrical resistance jump.

Tracking resistance measurements have been reported on graphite-reinforced epoxy composites [17].

Kalavtjidou et al. [74] measured the mechanical and electrical properties of polypropylene-based nanocomposites reinforced with up to 25 vol% exfoliated graphite nanoplatelets. The mechanical and electrical properties of the graphite nanocomposites were investigated and compared to the properties of polypropylene-based composites reinforced with other conductive materials, for example, carbon black and carbon fibers. It was found that the graphite nanoplatelets were the most effective at increasing the modulus of the polypropylene and comparable to the other materials in terms of percolation threshold.

Shen et al. [75] measured the electrical conductivity of polyethylene/maleic anhydride–grafted polyethylene/graphite nanocomposites. Electrical conductivity and morphology were influenced by the polymer preparation method and could be explained in terms of percolation theory.

5.8   EFFECT OF CARBON NANOTUBES
ON ELECTRICAL PROPERTIES OF POLYMER

Paul and Robeson [76] have reviewed the whole field of nanocomposites, including a detailed study of carbon nanotube reinforcement.

Nanotechnology is deeply embedded in the design of advanced devices for electronic and optoelectronic applications. The dimensional scale for electronic devices has now entered the nanorange. The utility of polymer-based nanocomposites in these areas is quite diverse, involving many potential applications, as well as types of nanocomposites. One specific nanocomposite type receiving considerable interest involves conjugated polymers and carbon nanotubes. A recent review of this area notes a litany of potential applications, including photovoltaic cells and photodiodes, supercapacitors, sensors, printable conductors, light-emitting diodes, and field effect transistors [77].

The electrical conductivity of carbon nanotubes in insulating polymers has also been a topic of considerable interest. The potential applications include electromagnetic interference shielding, transparent conductive coatings, electrostatic dissipation, supercapacitors, electromechanical actuators, and various electrode applications [78, 79]. The percolation threshold for electrical conductivity of epoxy composites containing multiwalled carbon nanotubes was found to be 0.0025 wt% multiwalled carbon nanotubes, considerably lower than nanoscale dispersed carbon black particles. The threshold conductivity of single-walled carbon nanotubes in epoxy composites was noted to be a function of the single-walled carbon nanotube type with values as low as 0.00005 vol fraction. Water-dispersed carbon black particles (42 nm) added to acrylic emulsions yielded electrical conductivity percolation levels as low as 1.5 vol% in dried films. In this system, the carbon black particles concentrate at the interference between the emulsion particles during drying, yielding a percolation network. The modulus of the emulsion system chosen as the matrix was noted to be an important variable with higher modulus (higher T_g), yielding lower threshold percolation values.

Carbon nanotubes or exfoliated graphite (graphene) offers substantial opportunities in the electrical/electronic/optoelectronics areas, as well as potential in specific emerging technologies. Carbon nanotube sheets have been proposed [80], and the potential for carbon nanotube–conjugated polymer composites would be of interest if sufficient electrical conductivity could be obtained (greater than 10^3 g/cm).

The incorporation of carbon nanotubes in various polymers as a means of improving electrical properties has been studied by various workers, as reviewed in Table 5.11.

Aoki et al. [99] used electrical conductivity and mechanical property measurements to evaluate the effect of carbon nanotube interface and dispersion uniformity on their properties.

Sung et al. [97] studied the electrical properties of polycarbonate-multiwalled carbon nanotubes. The multiwalled carbon nanotubes were functionalized by treatment with hydrogen peroxide. The hydrogen peroxide–treated multiwalled carbon

TABLE 5.11

Electrical Properties of Carbon Nanotube–Reinforced Polymers

Polymer	Electrical Property Measured	Reference
High-density polyethylene	Electrical conductivity	81
Polypropylene	Electrical properties	82
Epoxy resins	Electrical resistivity	83
Polyurethane	Electrical conductivity	84
Polylactic acid	Electrical conductivity	85
Polyphenylene vinylene	Electrical conductivity	86
Polyether ether ketone	Electrical conductivity	87
Polycarbonate	Electrical properties	88
Polyetherimide	Electrical properties	89
Polyimide	Electrical conductivity	90
Polylactide	Electrical properties	91
Multiwalled carbon nanotube-polypropylene	Electrical properties	92
Multiwalled carbon nanotube-high-density polyethylene	Electrical conductivity	93
Multiwalled carbon nanotube-polyimides	Electrical conductivity	94
Single-walled carbon nanotube-vinylene	Electrical properties	95
Carbon nanotube-polyether ether ketone	Electrical conductivity	96
Multiwalled carbon nanotube-polycarbonate	Electrical conductivity	97
T_iO_2-coated multiwalled carbon nanotube-epoxy composites	Volume resistivity	98
Carbon nanotube polyetherimide and epoxy resins	Electrical properties	99
Carbon nanotube polypyrrole	Electrical conductivity	100, 113, 114

nanotube was dried by thermal and freeze-drying methods. From the morphological studies, it could be seen that the degree of entanglement of the multiwalled carbon nanotubes was decreased after treating with hydrogen peroxide. For the hydrogen peroxide–treated multiwalled carbon nanotubes obtained by thermal drying, the length of the multiwalled carbon nanotubes was shorter than that of the hydrogen peroxide–treated multiwalled carbon nanotubes obtained by freeze drying. The rheological and electrical properties of the polycarbonate-multiwalled carbon nanotube (hydrogen peroxide–treated) composites increased compared to those of the untreated polycarbonate-multiwalled carbon nanotubes.

Also, the electrical conductivities showed higher values for the polycarbonate-multiwalled carbon nanotube (hydrogen peroxide treated with freeze drying) composites than the polycarbonate-multiwalled carbon nanotube (hydrogen peroxide treated with thermal drying) composites. From the results of the morphological, rheological, and electrical properties of the polycarbonate-multiwalled carbon nanotube composites, it was suggested that the electrical and rheological properties of the polycarbonate-multiwalled carbon nanotube composites are affected by the multiwalled carbon nanotube–multiwalled carbon nanotube network structure, which is related to the multiwalled carbon nanotube morphologies, such as the degree of agitation and the aspect ratio of the multiwalled carbon nanotubes.

5.9 EFFECT OF MONTMORILLONITE CLAY REINFORCEMENT ON ELECTRICAL PROPERTIES OF POLYMERS

Pandis et al. [101] examined the morphology, microhardness, and electrical properties of melt-mixed or compression-molded composites based on conductive polypyrrole dispersed into a nonconductive polypropylene matrix either as a pure component or as a nanocomposite with sodium montmorillonite. The effects of the polypyrrole and montmorillonite content on the properties of the composites were studied, and the results are discussed in terms of microstructure.

Ramar and Alagar [102] compared the electrical properties of organically modified nanoclay-reinforced ethylene-propylene-diene-y-tris(2-methyoxy-ethoxy) vinyl/silane-polystyrene and ethylene-propylene-diene-g-tris(2-methoxy-ethoxy)vinyl silane-polymethyl methacrylate blend.

These blends were characterized in different experimental studies to assess their suitability for cable insulation applications. The effects of incorporation of tris(2-methoxy-ethoxy)vinyl silane-ethylene-propylene-diene on electrical properties such as volume resistivity, surface resistivity, arc resistance, and dielectric strength of the blends were studied. The values of volume resistivity, surface resistivity, arc resistance, and dielectric strength were increased with increasing concentration of tris(2-methyl ethoxy)vinyl silane-ethylene-propylene-diene due to the presence of $-Si-O-Si$ linkages. The blends filled with nanoclay exhibited improved electrical properties due to the presence of inorganic moiety, which in turn enhanced the insulation behavior. Among the hybrid polystyrene and polymethyl methacrylate thermoplastic blends studied, however, ethylene-propylene diene-g-tris(2-methyl-ethoxy)vinyl silane-polystyrene blends possessed better insulating behavior than ethylene-propylene-diene-g-tris(2-methoxy-ethoxy)vinyl silane-polymethyl methacrylate blends due to thermally stable $-Si-O-Si$ linkages, hydrophobic and nonpolar in nature. X-ray diffraction studies were also carried out to confirm the formation of nanoclay-reinforced blends. Clay interaction was shown to occur in the case of polystyrene blends, whereas clay interaction with some degree of exfoliation occurred in the case of polymethyl methacrylate.

Lee et al. [103] studied the electrical and mechanical properties of nanocomposites derived from organophilic montmorillonite dispersed in polyvinyl chloride and poly(vinyl chloride-co-vinyl alcohol-co-vinyl acetate) matrices. They used a solution intercalation method in which montmorillonite was intercalated with dodecylamine through conversion of its amine (NHO_2'') groups into cations, which can form ionic bonds with the negatively charged silicate layers of montmorillonite. This organic clay was then dispersed in the solutions of polyvinyl chloride or its copolymer in various weight ratios.

Tsotra et al. [12] studied the electrical properties of clay-reinforced epoxy resin-polyaniline blends.

5.10   EFFECT OF ZINC OXIDE REINFORCEMENT ON ELECTRICAL PROPERTIES OF POLYMERS

Tjong et al. [16] showed that the dielectric constant of nanocomposites of annealed low-density polyethylene and zinc oxide was higher than that of quenched nano-composites, whereas the resistivity of annealed low-density polyethylene-zinc oxide nanocomposites was considerably lower than that of the quenched samples. The dielectric constant of both the annealed and quenched low-density polyvinyl-zinc oxide nanocomposites showed a pronounced frequency dependence when the zinc oxide volume content reached 52 vol% as a result of the formation of the zinc oxide network.

5.11   EFFECT OF MINERAL REINFORCEMENT IN ELECTRICAL PROPERTIES OF POLYMERS

The effect of the incorporation of minerals into polymer formulations is exemplified in Table 5.12, which lists the physical properties of virgin polydiallyl phthalate and the mineral-filled polymer. The only benefits of incorporating minerals are improvement in flow properties and a more uniform shrinkage, that is, lower warpage. These properties render the filled polymer suitable for applications such as relays, switches, contact bases, and communicators.

BID Ltd. has developed a new grade of mineral-filled polypropylene for use in electrical applications. Beetle PP120M and PP140M are mineral-filled grades that offer a 750°C glow wire rating with a high degree of dimensional stability and

TABLE 5.12
Physical Properties of Virgin Polydiallyl Phthalate and Mineral-Filled Polydiallyl Phthalate

	Virgin Polymer	Mineral-Filled Polymer
Maximum operating temperature, °C	160	160
Tensile strength, MPa	70	57
Flexural modulus, GPa	10.6	9.0
Elongation at break, %	0.9	0.9
Notched Izod impact strength, kj/m	0.41	0.35
Heat distribution temperature		
At 0.45 MPa°C	260	260+
At 1.8 MPa°C	210	190
Linear expansion coefficient, m/m/°C × 10^{-5}	3.0	3.2
Volume resistivity, log ohm-cm	14	15
Dielectric strength, Mv/m	16	15
Dielectric constant, 1 kHz	4.4	5.2
Dissipation factor, 1 kHz	0.01	0.04

quality of surface finish, while the higher-performance PP800 grade passes the glow wire test at 960°C and offers either a V0 or V2 flammability rating.

5.12 EFFECT OF SILICA REINFORCEMENT ON ELECTRICAL PROPERTIES OF POLYMER

Suwanpratech and Hattapanit [104] compared the use of black rice husk ash as a filler in epoxy resins used for embedding electrical and electronic devices to the use of commercial fillers, fused silica, and crystalline silica in various weight fractions between 20% and 60%. This led to an increased mixing viscosity, thermal expansion, and water absorption, with slightly lower tensile properties, compared to commercial silicon fillers. It was suggested that the use of different combustion conditions for the rice husk ash, to lower the carbon content, may make the results more comparable.

Li et al. [105] showed that in polymer-dispersed liquid films based on liquid crystals, a photopolymerizable acrylate matrix, and silica, the electro-optical properties could be controlled by the amount of silica nanoparticles.

5.13 EFFECT OF TALC REINFORCEMENT ON THE ELECTRICAL PROPERTIES OF POLYMERS

Saq'an et al. [106] determined the physical properties of a talc-filled polypropylene composite. Measurements included volume resistivity, electrical conductivity, and volume resistivity. The effect of talc content and temperature on the thermal, elastic, and electrical properties of the polymer was studied.

5.14 EFFECT OF IRON OXIDE AND HEMATITE NANOPARTICLES ON ELECTRICAL PROPERTIES OF POLYMERS

Iron oxide nanoparticles have been investigated for various applications, including electrical drug delivery, magnetic resonance imaging, contract enhancement, immunoassay, and cellular therapy. These investigations often employ magnetite (Fe_3O_4) dispersed in a polymeric microsphere or microcapsule involving biodegradable or natural polymers [107, 108].

Dudic et al. [109] studied the electrical properties of a composite comprising epoxy resins and alpha hematic rods. DC conductivity and DC current relaxation measurements showed a significant influence on Fe_2O_3 nanorods on the DC electrical properties of the epoxy matrix. However, the observed effects of the filler below and above the glass transition are different. Because of their high specific surfaces, nanorods affected the segmental mobility of epoxy molecules to a large extent, which resulted in an increase in the glass transition temperature (T_g) and a decrease in the dielectric permittivity in the high-frequency, low-temperature region. It was further observed that at elevated temperatures (above T_g) and low frequencies, the real part of dielectric permittivity of the nanocomposites exceeds that of the pure matrix; that is, there is a transition toward microcomposite-like dielectric behavior [57–120].

5.15 EFFECT OF MISCELLANEOUS REINFORCING AGENTS ON THE ELECTRICAL PROPERTIES OF POLYMERS

Flandin et al. [110] studied the influence of the degree of cure and the presence of inorganic fillers on the ultimate electrical properties of bisphenol A–based epoxy composites.

Niak and Mishra [111] studied the electrical properties of natural fiber high-density polyethylene composites.

Composites of banana, hemp, and agave with high-density polyethylene resin were separately prepared in different ratios: 60:40, 55:45, 50:50, and 45:55 (wt/wt). These fibers were also treated with maleic anhydride, and the effect of maleic anhydride on surface resistivity and volume resistivity of wood polymer composites was studied. The surface resistivity decreased with an increase in fiber content in the composites, while volume resistivity increased. The maximum surface resistivity and volume resistivity were observed in the untreated banana fiber composite, while minimum surface resistivity and volume resistivity were found in the maleic anhydride–treated agave fiber composite. The decrement in volume resistivity and surface resistivity was due to the increase in cross-linking between the polymer and fiber by treatment with maleic anhydride.

5.16 ELECTRICALLY CONDUCTIVE POLYMERS

Electrically conductive polymers are breakthrough materials that will open new markets and expand product ranges into new applications by offering materials with an expanded range of physical and electrical properties.

Electrically conductive polymers have applications in fields of electronics, microelectronic devices, computers, photography equipment, and other recent applications, such as electrically conducting electrochemical and fuel cells.

The measurement of the electrical conductivity of polymers is discussed in Section 5.3.8. Although, strictly speaking, unreinforced electrically conducting polymers are not within the subject area of this book, they are reviewed briefly here.

Conducting polymers were first produced in the mid-1970s as a novel generation of organic materials that have both electrical and optical properties similar to those of metals and inorganic semiconductors, but which also exhibit the attractive properties associated with conventional polymers, such as ease of synthesis and flexibility in processing.

Grimard et al. [112] have discussed various aspects of this subject:

1. Discovery of conducting polymers
2. Synthesis of conducting polymers
3. Conductivity and doping of conducting polymers
4. Use and modification of conducting polymers for biomedical applications, including general modification strategies for conducting polymers and biosensor applications of polypyrrole

The electrical conductivities of some doped polymers are listed in Table 5.13.

TABLE 5.13

Conductivity of Common Conducting Polymers

Conducting Polymer	Maximum Conductivity (S/cm²)	Type of Doping
Polyacetylene (PA)	200–1000	n, p
Polyparaphenylene (PPP)	500	n, p
Polyparaphenylene sulfide (PPS)	3–300	p
Polyparavinylene (PPv)	1–1000	p
Polypyrrole (PPy)	40–200	p
Polythiophene (PT)	10–100	p
Polyisothionaphthene (PITN)	1–50	p
Polyaniline (PANI)	5	n, p

Recent advances in carbon nanotube technology include the incorporation of carbon nanotubes into a number of conducting polymer-based biosensors. For example, preliminary studies have been performed exploring the properties of both polypyrrole-carbon nanotube and polyaniline-carbon nanotube devices as pH sensors [114]. One application used deoxyribonucleic acid–doped polypyrrole in conjunction with carbon nanotubes for detection of deoxyribonucleic acid.

In particular, the unique properties of polypyrrole-carbon nanotubes allowed the detection of hybridization reactions with complementary deoxyribonucleic acid sequences via a decrease in impedance [115]. Alternatively, similar deoxyribonucleic acid sensors have been created from a composite of polypyrrole and carbon nanotube functionalized with carbon groups to covalently immobilize deoxyribonucleic acid into carbon nanotubes [116, 117]. Carbon nanotubes have also been incorporated into biosensors as nanotube arrays into which enzymes can be immobilized, along with a conducting polymer [118] and a polypyrrole dopan [119]. In general, the presence of carbon nanotubes tends to increase the overall sensitivity and selectivity of biosensors.

Grimard et al. [112] concluded that conducting polymers, unlike many other materials, have uses in a diverse array of applications, ranging from photovoltaic devices to nerve regeneration. The unique property that ties all these applications together is the conductivity of the conductive polymer. Conductive polymers are organic in nature, making them more likely to be biocompatible; further, the presence of a conjugated backbone within the polymer endows it with the ability to conduct electrons, like metals/semiconductors, unlike any other polymer. In addition to these highly desirable properties, the ease of preparation and modification of conductive polymers has made them a popular choice for many applications. This is especially true in biomedicine, where many applications benefit from the presence of conductive materials, whether for biosensing or for control over cell proliferation and differentiation. Despite the vast amount of research already conducted on conductive polymers for biomedical applications, the field is still growing and many questions remain to be answered.

Fan et al. [121], for example, used in situ polymerization of pyrrole on carbon nanotubes to produce carbon nanotube-pyrrole composites. The electrical conductivity of the composite was shown to be appreciably higher than that of polypyrrole alone, and the composite exhibited semiconducting behavior over a wide range of temperatures.

Sag'an et al. [106] carried out electrical conductivity and Hall effect studies in order to investigate the nature, type, and development of charge carriers in conductive polymer composites containing polyaniline-based carbon fibers at different concentrations. The dependence of the electrical conductivity on temperature was characterized by a two-stage electrical conduction process with a semiconducting type of behavior and two activation energies. It was found that the measured Hall voltage varied linearly with the Hall current with two different signs of slopes. This suggested that a composite of low fiber content was functioning as p-type material, and then changed to n-type with increasing carbon fiber content to more than 15 wt%. The density of the charge carriers increased with carbon fiber content in a way similar to that of the electrical conductivity for all given composites, showing a percolation phenomenon. The calculated charge carrier density included the magnetostatic arising from both the polycarbonate matrix and the free charge carriers themselves. By considering the filled carbon fibers a random semiconducting material, the results obtained for various composites were described in terms of the band structure model. The composite bulk morphology was observed by scanning electron.

Tsotra et al. measured the electrical properties of clay-reinforced epoxy resin-polyaniline blends [12].

REFERENCES

1. ASTM D1401-02, Standard test method for water separability of petroleum oils and synthetic fluids, 2002.
2. DIN IEC 60093, Methods of test for insulating materials for electrical purposes: Volume resistivity and surface resistivity of solid electrical insulating materials, 1993.
3. DIN IEC 60167, Methods of test for insulating materials for electrical purposes: Insulating resistance of solid materials, 1993.
4. ISTN D257-99, Standard test method for DC resistance on conducting of insulating materials, 1999,
5. ASTM D150-98, Standard test methods for AC loss characteristics and permittivity (dielectric constant) of solid electrical insulation, 2004,
6. DIN 53483-1, Testing of insulating materials; determination of dielectric-properties; definitions, general information, 1969.
7. ASTM D149-97a, Standard test method for dielectric breakdown voltage and dielectric strength of solid electrical conductivity materials at commercial power frequencies, 2004.
8. ASTM D495-99, Standard test method for high voltage, low circuit, dry arc resistance of solid electrical insulations, 2004.
9. DIN IEC 60112, Method for the determination of the proof and the comparative tracking indices of solid insulating materials, 2004.
10. CEI 15-50, Guide per la Determinazione delle Proprieta' di Resistenza alla Sollecitazione Termica dei Materiali Isolanti Elettrici-Partes 2: Scelta dei Criteri di Prova, 1997.
11. UNI 4290, Tests on thermosetting plastics—Determination of resistance to tracking, 1959.
12. P. Tsotra, K.G. Garos, O. Grysthehub, and K. Friedrick, *Journal of Materials Science*, 2005, 40, 569.
13. J.B. Niak and S. Mishra, *Polymer Plastics Technology and Engineering*, 2005, 44, 687.

14. K. Tsuchiya, H. Ishii, Y. Shibasaka, S. Ando, and M. Mede, *Macromolecules*, 2004, 37, 4794.
15. Q.Q. Ke, X.-Y. Hunag, P. Wei, G.L. Wang, and P.K. Jiang, *Macromolecular Materials and Engineering*, 2006, 291, 1271.
16. S.C. Tjong, G.D. Liang, and S.P. Boa, *Journal of Applied Polymer Science*, 2006, 102, 1437.
17. C.-C. Wong, J.-F. Song, H.-M. Bao, Q.-D. Shen, and C.-Z. Yang, *Advanced Functional Material*, 2008, 18, 1299.
18. T. Seckin, S. Koytepe, N. Kivilcim, E. Bahce, and I. Adiguzel, *International Journal of Polymeric Materials*, 2008, 57, 429.
19. F.R. Diez, J. Moreno, L.H. Taylor, E.A. East, and D. Radic, *Synthetic Metals*, 199, 100, 187.
20. S.H. Hsiuo, C.A. Yang, and S.C. Huang, *Journal of Polymer Science, Part B: Polymer Chemistry*, 2004, 42, 2377.
21. C.P Yang, S.H. Hsiao, C.Y. Tsai, and G.S. Lio, *Journal of Polymer Science, Part B: Polymer Chemistry*, 2004, 42, 2416.
22. P. Kamezis, *Chemical Week*, 2002, 164, 24.
23. Axon Macplas International, May 2002, 66.
24. J. Seckin, S. Koytepe, N. Kivecleim, and I. Adiguzel, *International Journal of Polymer Materials*, 2008, 57, 429.
25. S.A. Cruz and M. Zanin, *Journal of Applied Polymer Science*, 2004, 91, 1730.
26. M. Nedjar, *Journal of Applied Polymer Science*, 2009, 11, 1985.
27. A.P. Matthew, V. Varyghese, and S. Thomas, *Journal of Applied Polymer Science*, 2005, 98, 2017.
28. P. Ramar and M. Alagar, *Progress in Rubber, Plastics and Recycling Technology*, 2008, 24, 121.
29. N. Parvatikar, S. Jain, S.V. Bhoraskar, and M.V.N. Abbika Prasad. *Journal of Applied Polymer Science*, 2005, 102, 5533.
30. D. Cyniak, J. Czekalski, and T. Jackoweski, *Fibres and Textiles in Eastern Europe*, 2005, 13, 16.
31. V.A. Beloshenko, V.N. Varyukhin, and Y.V. Yoznyak, *Composites, Part A: Applied Science and Manufacturing*, 2005, 361, 65.
32. L. Gao and X. Zhao, *Journal of Applied Polymer Science*, 2007, 104, 173.
33. F. Yakuhanoshi and M.E. Aydin, *Polymer Engineering and Science*, 2007, 47, 1016.
34. N.M. Mehta and P.H. Parsania, *Journal of Applied Polymer Science*, 2006, 100, 1754.
35. M. Drubeski, A. Siegmann, and M. Nankis, *Journal of Material Science*, 2007, 42, 1.
36. J. Suwanpratech and K. Hatthapanit, *Journal of Applied Polymer Science*, 2002, 86, 3013.
37. T. Sechin, S. Koytepe, N. Kiveclein, E. Bahoe, and T. Adiguzel, *International Journal of Polymer Materials*, 2008, 57, 421.
38. P. Tsotra and K. Freidrick, *Synthetic Metals*, 2004, 143, 237.
39. M. Singh, O. Odusanya, G.M. Wilmes, H.B. Eitouni, E.D. Gomez, A.J. Patel, V.L. Chen, M.J. Park, P. Fragauli, H. Iatrou, N. Hadjichistidis, D. Cookson, and N.P. Balsara, *Macromolecules*, 2007, 40, 4578.
40. A. Battin and C.E. Carraher, *Proceedings of ACS Polymeric Materials, Science and Engineering*, New Orleans, 2008, p. 396.
41. F. Yakuphanoglu and M.A. Schiavon, *Polymer Engineering and Science*, 2008, 48, 837.
42. Y.S. Lin and S.S. Chiu, *Journal of Adhesion Science and Technology*, 2008, 22, 1673.
43. A.W. Rinaldi, R. Matos, A.F. Rubira, O.P. Ferreira, and E.M. Giratto, *Journal of Applied Polymer Science*, 2005, 96, 1710.
44. C. Severance and D. Nobbs, *Proceedings of 64th Annual SPE ANTEC Conference*, Charlottesville, VA, 2006, p. 471.

45. V.M. Yoham, I. Raja, P.B. Bhargav, A.K. Sharma, and B.V.R.N. Rao, *Journal of Polymer Research*, 2007, 14, 283.
46. D.T. Seshadri and N.K. Bhat, *Journal of Polymer Science, Part B: Polymer Physics*, 2007, 45, 1127.
47. G.V.S. Reddy, A.K. Sharma, and B.V.R.N. Rao, *Polymer*, 2006, 47, 1318.
48. M. Singh, O. Odusanya, G.M. Wilmes, H.Z. Eitouni, E.D. Gomez, A.J. Patel, L. Chen, and P. Fragonli, *Macromolecules*, 2007, 40, 4580.
49. K. Khalid and F. Mohammed, *Express Polymer Letters*, 2007, 1, 711.
50. G. Calmak, E. Kneukyavuz, Z. Kucayavuz, and H. Lakmak, *Composites, Part A: Applied Science and Manufacturing*, 2004, 35, 417.
51. Y. Li, S. Wang, Y. Zhang, and Y. Zhang, *Journal of Applied Polymer Science*, 2005, 98, 1145.
52. T. Nishikawa, K. Ogi, T. Tana, Y. Okono, and I. Taketa, *Advanced Composite Materials*, 2007, 16, 1.
53. L. Sungho, K. Myung-Soo, and A.A. Ogale, *Proceedings of 62nd SPE Annual Conference*, Chicago, May 16–20, 2004, p. 1652.
54. S. Tjong, S. Liang, and S. Bao, *Polymer Engineering Science*, 2008, 48, 177.
55. W.H. Di, G. Zhang, Z.D. Zhao, and Y. Peng, *Polymer International*, 2004, 53, 449.
56. F. Fourrier and L. Pren, *Revista de Plasticos Modernos*, 2000, 29, 729.
57. S. Sagan, A.M. Zihlif, S.R. Al-Ani, and G. Ragasta, *Journal of Materials Science, Materials in Electronics*, 2008, 19, 1079.
58. S. Saq'an, A.M. Zihlif, R.S. Arl-Ani, and G. Ragosla, *Journal of Materials Science, Materials in Electronics*, 2007, 18, 1203.
59. A.N. Ogata, *Proceedings of 62nd SPE Annual Conference (ANTEC 2004)*, Chicago, May 16–20, 2004, p. 652.
60. V.M. Mahon, V. Raja, R.B.R. Langav, A.K. Sharmc, and V.V.R.N. Rao, *Journal of Polymer Research*, 2007, 14, 283.
61. E. Knickerboc, *Proceedings of SPE Annual Conference (ANTEC 2007)*, Cincinnati, OH, May 6–11, 2007, p. 1464.
62. M.C. Pandis, E. Logakis, V. Peoglas, M. Omastooa, M. Mravakov, A. Janki, J. Pointeck, Y. Penev, and L. Minkova, *Journal of Polymer Science, Part B: Polymer Physics*, 2009, 47, 407.
63. G. Camake, Z. Kukukyavuz, S. Kuccikyavuz, and H. Cakmak, *Composites, Part A: Applied Science and Manufacturing*, 2004, 35A, 417.
64. P. Tsetra and K. Freidrick, *Synthetic Metals*, 2004, 143, 237.
65. T. Tantaway, K. Kamade, and H. Ohnabe, *Journal of Applied Polymer Science*, 2003, 87, 97.
66. B. Jurkawska, B. Jurkawski, R. Ramrovski, S.S. Resetski, V.N. Kovel, L.S. Pinchuk, and Y.A. Olkhov, *Journal of Applied Polymer Science*, 2006, 100, 390.
67. P. Zhou, W. Yu, C. Zhou, F. Lia, L. Hou, and J. Wang, *Journal of Applied Polymer Science*, 2007, 103, 487.
68. X.-Y. Yin, B.X. Wang, and D.X. Chan, *Journal of Functional Polymers*, 2007, 19, 133.
69. Y. Li, S. Wang, Y. Zhang, and Y. Zhang, *Polymers and Polymer Composites*, 2006, 14, 377.
70. B. Pourabas and J. Peyghambardoost, *Proceedings of the Symposium on Polymers for Advanced Technologies*, Budapest, Hungary, September 13–16, 2005, paper 48, p. 2. Available from http//www.e-polymers.org/PAT2005epolymers/proceedings.htm.
71. Z. Chen, L. Christiansen, and J.R. Dahn, *Journal of Applied Polymer Science*, 2004, 91, 2949.
72. Y. Li, S. Wang, and Y. Zhang, *Journal of Applied Polymer Science*, 2005, 98, 1142.
73. M. Khelid and F. Mohammed, *Express Polymer Letters*, 2007, 1, 711.

74. K. Kalavtzidou, H. Fukushia, and L.T. Orzell, *Proceedings of 62nd SPE Annual Conference (ANTEC 2004)*, Chicago, May 16–20, 2004, p. 1533.

75. J.W. Shen, H.Y. Huang, S.W. Zuo, and J.H. Han, *Journal of Applied Polymer Science*, 2005, 97, 51.

76. D.R. Paul and L.M. Robeson, *Polymer*, 2008, 49, 3187.

77. M. Baibarac and P.J. Gómez-Romero, *Nanoscience and Nanotechnology*, 2006, 6, 1.

78. R.H. Baughaman, A.A. Zakhidov, and W.A. De Heer, *Science*, 2002, 297, 787.

79. M. Moniruzzaman and K.I. Winey, *Macromolecules*, 2006, 39, 5194.

80. M. Zhang, S. Fang, A.A. Zakhidov, S.B. Lee, A.E. Aliev, and C.D. Williams, *Science*, 2005, 309, 1215.

81. B. Plage and H.R. Schulten, *Journal of Polymer Science*, 1988, 21, 2018.

82. A. Ballisteri, D. Garozzo, G. Montaude, A. Policino, and M. Guiffrida, *Polymer*, 1981, 28, 139.

83. Z. Wang, G. Wa, Y. Hu, Y. Ding, H. Ha, and W. Fan, *Polymer Degradation and Stability*, 2002, 77, 427.

84. V.V. Zuev, F. Bertini, and G. Audisio, *Polymer Degradation and Stability*, 2000, 69, 169.

85. V.V. Zuev, F. Bertini, and G. Audisio, *Polymer Degradation and Stability*, 2002, 71, 213.

86. J.K. Haken and L. Tan, *Journal of Polymer Science*, 1987, 25, 1451.

87. X.H. Li, Y.Z. Meng, Q. Zhu, and S.C. Tjong, *Polymer Degradation and Stability*, 2003, 81, 157.

88. R.S. Lehrle and C.S. Pattenden, *Polymer Degradation and Stability*, 1998, 62, 211.

89. D. Yang, S.D. Li, W.W. Fu, J.P. Zhong, and D.M. Jia, *Journal of Applied Polymer Science*, 2002, 87, 199.

90. L. Ren, W. Fu, Y. Hao, H. Hu, D. Jia, and J. Shen, *Journal of Applied Polymer Science*, 2003, 91, 2295.

91. S.K. Ghosh, D. Repp, D. Khestgir, and S.K. De, *Proceedings of 151st ACS Rubber Division Meeting*, Anaheim, CA, 1997, paper 93.

92. S.H. Lee, M.W. Kim, S.H. Kim, and J.R. Joun, *European Polymer Journal*, 2008, 44, 1620.

93. X.L. Chen, H.Y. Fu C. Zhang, and S.Y. Zhu, *Polymer Materials Science and Engineering*, 2008, 24, 87.

94. J.N. Yuen, E.C.M. Ma, C.L. Chiang, C.C. Teng, and Y.H. Yu, *Journal of Polymer Science, Part B: Polymer Chemistry*, 2008, 46, 803.

95. H. Aarah, M. Baitonl, J. Wery, R. Almairac, S. Lefrant, E. Faulgues, J.L. Duval, and M. Hamedou, *Synthetic Metals*, 2005, 155, 63.

96. D.S. Bangarysampath, V.A. Altstardi, H. Ruckdreschel, J.K.W. Sandler, and M.S.P. Shaffer, *Journal of Plastics Technology*, 2008, 6, 19.

97. Y.T. Sung, M.S. Han, K.H. Song, J.W. Jung, H. Lee, L.K. Kum, J. Joo, and W.N. Kim, *Polymer*, 2006, 47, 4434.

98. S.M. Yuen, C.C.M. Ma, C.Y. Chuang, Y.H. Hsigo, C.L. Chiang, and A.D. Yu, *Composites, Part A*, 2008, 39, 119.

99. T. Aoki, Y. Ono, and T.O. Ogasawara, *Proceedings of the Twenty-First Technology Conference of the American Society for Composites*, Dearborn, MI, September 20, 2006.

100. J. Fan, M. Wan, D. Zhu, and B. Chang, *Synthetic Metals*, 1999, 102, 1266.

101. C. Pandis, E. Logaskis, V. Reoglas, P. Pissis, M. Onastova, M. Mravcakova, A. Janke, J. Piontek, Y. Peneva, and L. Minkova, *Journal of Polymer Science, Part B: Polymer Physics*, 2009, 47, 407.

102. P. Ramar and M. Alagar, *Progress in Rubber, Plastics and Recycling Technology*, 2008, 24, 121.

103. J.W. Lee, S.Y. Kim, and S.K. Khim, *Journal of Materials Science*, 2007, 42, 2486.

104. J. Suwanratech and K. Hattapanit, *Journal of Applied Polymer Science*, 2002, 86, 3013.

105. W. Li, M. Zhu, X. Ding, B. Li, W. Huang, H. Cao, Z. Yang, and H. Yang, *Journal of Applied Polymer Science*, 2009, 111, 1449.

106. S. Sagian, A.M. Zahtif, and G. Ragosta, *Journal of Thermoplastic Composite Materials*, 2008, 21, 457.

107. M. Gupta and A.K. Gupta, *Biomaterials*, 2005, 26, 3995.

108. C.C. Berry, *Journal of Materials Chemistry*, 2005, 15, 543.

109. D. Dudic, M. Marinovic-Cincovic, J.M. Nedeljkovic, and V. Djokovic, *Polymer*, 2008, 49, 4000.

110. L. Flandin, L. Vouyovtch, A. Beroual, J. Besrede and N.D. Alberola, *Journal of Physics*, 2005, 38, 144

111. J.B. Niak and S. Mishra, *Polymer Plastics Technology and Engineering*, 2005, 44, 687.

112. N.K. Grimard, N. Gomez, and C.E. Schmidt, *Progress in Polymer Science*, 2007, 32, 876.

113. H. Castano, E.A. O'Rear, P.S. McFeridge, and V.F. Sikavistas, *Macromolecular Bioscience*, 2004, 4, 785.

114. N. Ferrer-Anglada, M. Kaempgen, and S. Roth, *Phys Status Solidi B*, 2006, 243, 3519.

115. H. Cai, Y. Xu, P.G. He, and Y.Z. Fang, *Electroanalysis*, 2003, 15, 1864.

116. G. Cheng, J. Zhai, Y. Tu, P. He, and Y.A. Fang, *Analytica Chimica Acta*, 2005, 533, 11.

117. Y. Xu, X. Ye, L. Yang, P. He, and Y. Fang, *Electroanalysis*, 2006, 18, 1471.

118. L. Qu, P. He, L. Li, M. Goa, G. Wallace, and L. Dai, *Quantum Sensing and Nanophotonic Devices II Proceedings*, San Jose, CA, January 23–27, 2005, p. 82.

119. J. Wang and M. Musameh, *Analytica Chimica Acta*, 2005, 539, 209.

120. Y.-C. Tsai, S.-C. Li, and S.-W. Liao, *Biosensors Bioelectronic*, 2006, 22, 495.

121. J.H. Fan, M. Wan, D. Zhu, B. Chang, Z. Wei, and P.S. Xie, *Synthetic Metals*, 1991, 102, 1266.

6 Thermal and Thermooxidative Degradation of Reinforced Polymers

The degradation of filled polymers is characterized by a number of features that are atypical of normal (unfilled) polymers. The degradation of polymers proceeding at high temperatures in vacuo or in an inert gas atmosphere (helium, argon, etc.) is referred to as thermal degradation, while in ambient air or oxygen, the term is thermal-oxidative degradation. However, these features are normally associated with the prehistory of production of the filled polymer. In particular, the methods available for the introduction of fillers may influence not only the physical and chemical properties of polymers, but also their molecular characteristics. Thus, the mixing of melts or solutions of polymers with disperse fillers causes, in a number of cases, noticeable shifts in their molecular mass distribution. This is mainly associated with the mechanocracking of filled polymers, which is enhanced in the presence of fillers. The fragments of macromolecules produced interact either with each other or with the filler surface to form a grafted layer. These mechanochemical processes leading to variations in the molecular characteristics of polymers also affect thermal and thermal-oxidative stability, normally decreasing it. In addition, the conditions of introduction of fillers into polymers (temperature, concentration, intensity of mixing, environment, presence of adsorbed moisture, oxygen, etc.) also have a substantial effect on the polymer decomposition process. The filler plays a key role in the decomposition processes of polymer fillers, since the latter affect the whole complex of properties and structures of the polymers produced.

The experimental results of studies on the degradation of filled polymers are often contradictory. This may be caused not only by differences in the methods of production of the filled polymers, but also by the apparatus and experimental procedures (measurements of temperature, pressure, changes in the mass of samples, analysis of low- and high-molecular-mass products of degradation, etc.). Consequently, the description of the apparatus and methods for studies of the degradation of polymers, especially of filled ones, employed in different laboratories is essential, since it may promote the elimination of contradictions and errors that have been observed in certain investigations.

A wide range of techniques have been applied to the elucidation of the thermal stability of polymers:

Thermal Degradation	Thermooxidative Degradation
Thermogravimetric analysis	Thermogravimetric analysis [1, 2]
Differential thermal analysis	Differential scanning calorimetry [3–10]
Differential scanning calorimetry	Pressure differential scanning calorimetry
Thermal volatiliszation analysis	Evolved gas analysis
Evolved gas analysis	Infrared spectroscopy [11–25]
Mass spectroscopy methods	Electron spin resonance spectroscopy [26–31]
Matrix-assisted laser desorption/ionization	Mass spectrometry
Imaging chemiluminescence	Pyrolysis-based techniques

Thermal analysis methods can be broadly defined as analytical techniques that study the behavior of materials as a function of temperature [33]. These are rapidly expanding in breadth (number of thermal analysis–associated techniques) and depth (increased applications). Conventional thermal analysis techniques include differential scanning calorimetrical, differential thermal and thermogravimetric, thermomechanical, and dynamic mechanical analyses. The thermal analysis of a material can be destructive or nondestructive, but in almost all cases, subtle and dramatic changes accompany the introduction of thermal energy. Thermal analysis can offer advantages over other analytical techniques, including variability with respect to application of thermal energy (stepwise, cyclic, continuous); small sample size; the material can be in any solid form—gel, liquid, glass, or solid; ease of variability and control of sample preparations; and ease and variability of atmosphere. It is relatively rapid, and instrumentation is moderately priced. Most often, thermal analysis dates are used in conjunction with results from other techniques.

Cerrada et al. [34] reviewed the current trends in the thermal testing of polymers. Some of the thermal and chemical stability applications to which these various techniques have been applied are summarized in Table 6.1.

6.1   GLASS FIBER REINFORCEMENT

The effect that reinforcing agents can have on the decomposition profiles of a polymer is illustrated in Figure 6.1, which shows decomposition profiles for fiberglass-reinforced polytetrafluoroethylene, which is seen to start to degrade at 350°C.

Pasquale et al. [35] applied thermogravimetric analysis and online flash pyrloysis–gas chromatography–mass spectrometry in a study of the thermal decomposition of polyamides.

Prethermal decomposition of glass fiber–reinforced and virgin polyamide started at 350°C and proceeded with a weight loss of 100% under nitrogen for the unreinforced polymers [36–39]. Predominant decomposition products were cyclopentanone and epsilon-caprolactam.

TABLE 6.1
Techniques in Thermal and Chemical Stability Studies

Analysis Method	Thermal Stability	Chemical Stability
TGA	Weight-loss measurement	Chemical stability
	Moisture content	Chemical composition
	Thermal stability	Decomposition kinetics
	Quality control testing	Catalyst activity
DTA	Thermal stability	Chemical stability
	Quality control testing	Chemical composition
		Decomposition kinetics
		Extent of rate of resin cure
		Catalyst activity
DSC	Thermal stability	Chemical stability
	Quality control testing	Chemical composition
		Decomposition kinetics
		Extent of rate of resin cure
		Resin cure kinetics
		Catalyst activity
TVA	—	Polymer decomposition
		Identification of volatiles
EGA	—	Polymer decomposition
		Identification of volatiles

Note: TGA = thermogravimetric analysis, DTA = differential thermal analysis, EGA = evolved gas analysis

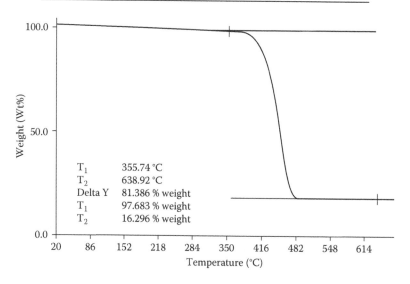

T_1	355.74 °C
T_2	638.92 °C
Delta Y	81.386 % weight
T_1	97.683 % weight
T_2	16.296 % weight

FIGURE 6.1 TGA decomposition profiles of polyamide.

Glass fibers also cause thermal degradation to occur in polymethyl methacrylate [36, 37, 40].

6.2   CLAY REINFORCEMENT

Wong et al. [41] showed that sodium montmorillonite imparted a better thermal stability to polyvinylidene difluoride-polyethylene (PE) glycol polymers than did organic montmorillonite. They showed that ethyl-3-methylimidazolium tetrafluoro-borate–functionalized montmorillonite greatly enhanced the thermal stability of the polymer. Liu et al. [42] studied the effect of various organoclays on the heat stability of polytrimethylene terephthalate.

Ardhnananta et al. [43] showed that organoclay improved the thermal stability of ethylene vinyl acetate nanocomposites.

Stoehler et al. [44] studied the thermal properties of polyethylene-montmorillonite nanocomposites and found evidence for accelerated formation and decomposition of hydroperoxides during the thermooxidative degradation of the nanocomposites in the range of 170°C to 200°C, as compared to unfilled polyethylene, as well as the formation of intermolecular chemical cross-links in the nanocomposites above 200°C due to recombination reactions involving the radical products of hydro-peroxide decomposition.

The reinforcing properties of organically modified montmorillonite clay have been studied in several other polymers, including liquid crystalline polymers [45], and the suspension of montmorillonite in aqueous media containing polyelectrolytes [46].

6.3   SILICA REINFORCEMENT

The dates in Figure 6.2 show that the incorporation of silica in amounts between 10% and 50% has little effect on the degradation profile of polytetrafluoroethylene when compared with the curve obtained for unreinforced polytetrafluoroethylene.

Gebe and Runt [47] studied the effects of organically modified layered silicates, such as trioctyl methyl ammonium–modified smetite clays or octadecyl ammonium–modified silicate, on the thermal properties of a 75% polyvinylidene fluoride-*co*-trifluoroethylene random copolymer.

Silica did not have any adverse effect on the thermal stability of polyethylene [48] and polystyrene [49–51]; indeed, it has been claimed that silica improves the thermal stability of these polymers. This has been proved by thermogravimetric analysis, which indicated that thermal stability was improved in nanocomposites with higher organically modified silicate contents. Differential scanning calorimetry results showed that thermal transitions in the nanocomposites depended on organi-cally modified silicate content. Nanocomposites with 2% organically modified sili-cate exhibited a crystal nucleate effect, giving a significant increased in the amount of ferroelectric crystals formed during cooling. For 10%–25% organically modified silicate, the amounts of paraelectric and ferroelectric crystals were reduced. The larger organically modified silicate additions depressed the melt-to-paraelectric tran-sition temperature, while an increase in the paraelectric-to-ferroelectric transition

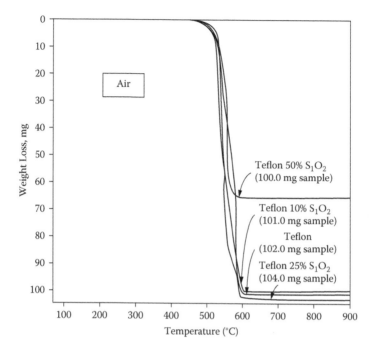

FIGURE 6.2 Degradation profiles of virgin polytetrafluoroethylene and polytetrafluoro-ethylene containing between 10% and 50% of silica.

temperature was observed for all compositions. Upon reheating, the ferroelectric phase transition showed significant hysteresis.

In Nanomer OMS, weight loss, that is, decomposition, begins at a lower temperature than in the copolymer, which is due to the loss of OMS surfactants.

Assuming that the surfactant is removed by the time the temperature reaches 550°C, its amount can be estimated from the weight loss. Table 6.2 reports the weight loss of polytetrafluoroethylene from 50 to 550°C with lucentite and nanomer OMS losing, respectfully, 29% and 30% of weight over this temperature range.

This translates into lower initiation of weight loss in the nanocomposites. However, the overall thermal stability of the composites at very high temperature is significantly improved by the addition of organically modified silicate.

6.4   CARBON NANOTUBE REINFORCEMENT

Lai et al. [52] and Loos et al. [53] demonstrated a great improvement in mechanical properties and a slight improvement in thermal properties following the addition of carbon nanotubes to epoxy resins.

Kim et al. [54] studied the thermal and electrical properties of poly(L-lactide)–grafted multiwalled carbon nanotube composites using thermogravimetric analysis. They found that poly(L-lactide)–grafted composites made with multiwalled carbon nanotubes have a higher activation energy and, consequently, a higher thermal stability than poly(L-lactide)–grafted multiwalled carbon nanotubes.

TABLE 6.2

Thermogravimetric Analysis: Relative Weight Loss at 550°C and Residue at 1000°C for Polyvinylidene Fluoride-Co-Trifluoroethylene Random Copolymers, OMS Values

Sample	OMS (%)	% Wt Loss 50°C–550°C		% Residue at 1000°C	
		Expected	Observed	Expected	Observed
75/25[a]	0	89	89	5.2	5.2
N	25	75.1	60.4	19.4	22.6
N	100	33.7	33.7	62.1	62.1
L	10	83.2	71.9	10.8	6.8
L	25	74.5	60.2	19.3	32.3
L	100	29.9	29.9	62.7	62.7

[a] 75/25 refers to copolymer; N refers to Nanomer OMS; L refers to Lucentite. OMS improved OMS addition. This table compares the thermogravimetric analysis results of composites based on 75:25 polyvinylidene fluoride-*co*-trifluoroethylene random copolymer-lacentite OMS (L) and 75:25 polyvinylidene fluoride-*co*-trifluoroethylene random copolymer-Nanomer OMS (N).

Comparisons of thermal properties of polymer before and after reinforcement with multiwalled carbon nanotubes have also been reported for polyhydroxy butyrate-*co*-hydroxy valerate [52] and polycaprolactomy composites [55].

6.5　TALC REINFORCEMENT

Talc causes thermal degradation of polyethylene [56–68, 69], reduces thermal degradation of polyvinyl acetate [65, 75, 76], and has no effect on the thermal stability of polymethyl methacrylate [66, 68, 69].

6.6　CARBON FIBER CARBON AND GRAPHITE REINFORCEMENT

Carbon fiber causes thermal degradation of polyamides at elevated temperatures [70]. Carbon does not affect the thermal stability of polymethyl methacrylates [70, 71] or polystyrene [49–51], but reduces the thermal stability of polyvinyl acetate [72, 73]. Graphite improves the thermal [49, 64, 65] stability of polystyrene and polyamides [38, 39, 74, 75].

6.7　MINERAL REINFORCEMENT

Minerals reduce the thermal stability of vinyl acetate [65, 73] and polymethyl methacrylate [104].

Kaolin, however, does cause degradation of polyethylene at elevated temperatures [56, 76, 77].

6.8 REINFORCEMENT WITH OTHER COMPOUNDS

Other reinforcing agents either have been used to reduce the thermal degradation of polymers or do not have an effect on it. These include calcium carbonate in polyvinyl acetate [65, 73] and polymethyl methacrylate [69, 71] and titanium dioxide-magnesium oxide in polyethylene [78].

6.9 EFFECT OF METALS ON THE HEAT STABILITY OF POLYMERS

Experimental work has been carried out on the effect of metals on thermooxidative degradation of reinforced plastics.

Various tests have been employed to follow the thermal oxidative degradation. These include the measurement of the rate of oxygen uptake by the specimen, changes in molecular weight, changes in impact strength, and changes in flexural endurance, that is, flex testing. The rate of onset of thermal degradation is influenced by the rate of diffusion of oxygen into the specimen via the surface, as well as the chemical kinetics of the reactions between oxygen and the polymer. Consequently, in such testing, sample thickness must be carefully controlled. One measure of polymer degradation due to the effect of metals is that of the time at which 50% of the exposed specimens fail in a standard flex test.

Test results are conveniently presented as a graph of exposure temperature on a linear scale versus the time for 50% failure on a logarithm scale. It is shown in Figure 6.3, for example, that the metal having the most deleterious effect on the heat aging life of polypropylene is copper, followed by brass, nickel, and chromium, not in order of severity. Nickel and chromium reduce the heat aging life by three- and

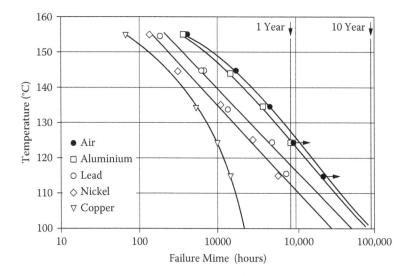

FIGURE 6.3 Heat aging of a stabilized polypropylene in air and in contact with aluminum, lead, nickel, and copper.

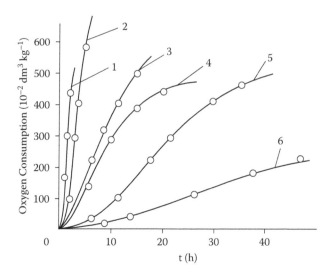

FIGURE 6.4 Kinetics of oxygen consumption of polyethylene containing 10^{-7} to 10^{-3} mol/mol PE of cobalt stearate (1), cobalt chloride (2), cobalt oxide Co_3O_4 (3), cobalt sulfate $CoSO_4$ (4), metallic cobalt (5), and PE without filler (6).

twofold, respectively, when compared to results obtained in virgin polypropylene aluminum and tine, having a barely detectable effect on aging.

The possibility of interactions between the heat-stable additives and other additives employed cannot be ignored. For example, most applications require pigmented material, so it was considered necessary to study the effect (if any) of the pigment systems used in the polymer on the heat stability of the polymer. Because most pigments used to color polymers are based on inorganic compounds (e.g., cadmium sulfides, zinc sulfides, sulfoselenides) and because certain metals are known to catalyze the thermal degradation of polymer pigments, such pigments may accelerate this degradation at elevated temperature.

Figure 6.4 shows a comparison of oxygen consumption (in 10^{-2} dm³ kg⁻¹) obtained for virgin polypropylene (curve 6) with the much higher oxygen consumption obtained with polyethylene containing increasing amounts of cobalt salts (curves 1–5).

Introduction into polymethyl methacrylate of disperse zinc or aluminum at the point of polymerization leads to a marked stabilization of the polymer filled only with zinc (Figure 6.4, curves 4–6), on which effect related to the nonchain inhibition of the thermal-oxidative degradation due to breaking of the oxygen bond by unoxidized particles of zinc. Since aluminum particles are covered with thick layers of oxide film, they exert practically no stabilizing influence on the thermal-oxidative degradation of the polymer (Figure 6.4, curves 1 and 2).

Table 6.3 illustrates the effect on thermooxidative stability of polyethylene filled with highly dispersed iron (expressed as the temperature at which a given weight loss between 2% and 50% occurs).

It has been seen that in each case, the stability of polymer containing up to 1% iron is greater than that in the case of virgin polyethylene.

TABLE 6.3

Thermal-Oxidative Stability of Polyethylene Filled with Highly Disperse Iron

Composition	Temperature at Which a Given Weight Loss Occurs (K)		
	2%	10%	50%
Initial PE	483	578	675
PE + 0.015% Irganox 1010	543	633	713
PE + 0.05% Fe	523	613	713
PE + 0.10% Fe	553	633	723
PE + 0.25% Fe	563	653	713
PE + 0.50% Fe	573	653	713
PE + 1.00% Fe	583	683	713

Generally speaking, and with some exceptions, metals have an influence on the thermal stability of polymers, as reviewed below:

Polyethylene: Aluminum oxide, ferric oxide [56–59, 79–87], copper oxide [48], and iron [57–63] all thermally degrade the polymer, whereas highly dispersed iron, copper, lead [60], and cobalt chloride [48] have no effect on polymer stability at deviated temperatures.

Polydiene: Copper, cupric oxide, cuprous oxide, manganese, and cobalt are catalysts for the thermal degradation of this polymer [88–91], whereas iron, nickel, zinc, and lead do not cause polymer degradation [88–91].

Polymethyl methacrylate: Copper, cadmium, and aluminum coenzyme inhibit but do not prevent thermal degradation; barium calcium and strontium oxides affect the thermal behavior of the polymer [71].

Polyvinyl chloride: Zinc oxide, copper oxide alumina, and ferric oxide all adversely affect thermal stability [92], whereas copper improves the thermal stability [93].

Ethylene-vinyl acetate: Magnesium hydroxide aluminum–layered double hydroxide nanocomposites are thermally unstable [94].

Phenol formaldehyde: Resins, nickel, copper, and cobalt [95–97] all cause thermal decomposition of polymers.

Polyphenylene sulfone: Nickel, aluminum, copper, silver, and alumina all adversely affect the thermooxidative stability of the polymer, whereas tin does not [98–100].

Polyethylene terephthalate: Iron oxide causes polymer degradation of the polymer [40].

Polyorgano giloxanes: Copper, lead, nickel, zinc, iron, and cadmium cause thermal degradation of the polymer-elevated temperature [101–110].

Polyurethanes: Lead and lead oxides accelerate thermal degradation of this polymer.

REFERENCES

1. M. Thizon, C. Eon, P. Valentin, and G. Guichon, *Analytical Chemistry*, 1976, 48, 1861.
2. P.M. Norling and A.V. Taboesky, in R.T. Conley (ed.), *Thermal Stability of Polymers*, Marcel Dekker, New York, 1970, chap. 5.
3. A.A. Duswalt, *Thermochimica Acta*, 1974, 8, 57.
4. T. Zaharescu, V. Meltser, and R. Vilcu, *Polymer Degradation and Stability*, 1999, 64, 101.
5. L. Woo, C.L. Sandford, and S.Y. Ding, *Polymer Preprints*, 2001, 42, 394.
6. M. Groening and M. Hakkarainen, *Journal of Applied Polymer Science*, 2002, 86, 3396.
7. S. Ding, M.K.T. Ling, A.R. Khare, C.L. Sandford, and L. Woo, *Journal of Applied Medical Polymers*, 2000, 4, 28.
8. R.G. Scholz, J. Bednarczyk, and T. Yamuchi, *Analytical Chemistry*, 1966, 38, 331.
9. J.C.W. Chien and J.K.L. Kiang, *Makromolecular Chemie*, 1980, 181, 47.
10. T.P. Wampler and E.J. Levy, *Journal of Analytical and Applied Pyrolysis*, 1985, 8, 153.
11. L.H. Cross, R.B. Richards, and H.A. Willis, *Discussions of the Faraday Society*, 1950, 9, 235.
12. F.M. Rugg, J.J. Smith, and R.C. Bacon, *Journal of Polymer Science*, 1954, 13, 535.
13. A.W. Pross and R.M. Black, *Journal of the Society of Chemical Industry* (London), 1950, 69, 115.
14. H.C. Beachell and S.P. Nemphos, *Journal of Polymer Science*, 1956, 21, 113.
15. H.C. Beachell and G.W. Tarbet, *Journal of Polymer Science*, 1960, 45, 451.
16. J.P. Luongo, *Journal of Polymer Science*, 1960, 42, 139.
17. G.D. Cooper and M. Probe, *Journal of Polymer Science*, 1960, 44, 397.
18. G. Kimmerle, *Combustion Toxicology*, 1974, 1, 4.
19. H. Kveder and G. Ungar, *Nafta (Zagreb)*, 1973, 24, 85.
20. Yu A. Pentin, B.N. Tarasevich, and B.N. El'tsefon, *Vestnick-Moskovskii Univerrsitet Khimia*, 1973, 14, 13.
21. D.L. Tabb, J.J. Sevcik, and J.L. Koenig, *Journal of Polymer Science, Part B: Polymer Physics*, 1975, 13, 815.
22. P.J. Miller, J.F. Jackson, and R.S. Porter, *Journal of Polymer Science, Part B: Polymer Physics*, 1973, 11, 2001.
23. M.G. Chan and D.L. Allara, *Polymer Engineering Science*, 1974, 14, 12.
24. N. Grassie and N.A. Weir, *Journal of Applied Polymer Science*, 1965, 9, 963.
25. V. Grassie and N.A. Weir, *Journal of Applied Polymer Science*, 1965, 9, 975.
26. S.-I. OhnishI, S.-I. Sugimoto, and I. Nitta, *Journal of Polymer Science, Part A: General Papers*, 1963, 1, 625.
27. R. Salovey and W.A. Yager, *Journal of Polymer Science, Part A: General Papers*, 1964, 2, 219.
28. Y. Hajimoto, N. Tamura, and S. Okamoto, *Journal of Polymer Science, Part A: General Papers*, 1965, 3, 255.
29. I. Ouchi, *Journal of Polymer Science, Part A: General Papers*, 1965, 3, 2685.
30. Y. Sakai and M. Iwasaki, *Journal of Polymer Science, Part A: Polymer Chemistry*, 1969, 7, 1749.
31. Y. Hama and K. Shinohara, *Journal of Polymer Science, Part B: Polymer Physics*, 1970, 8, 651.
32. G. Gallet, S. Carroccio, P. Rizzarelli, and S. Karlsson, *Polymer*, 2002, 43, 1081.
33. P. Cabe, M. Jaffe, and C.W. Carraher, *Proceedings of ACS Polymer Materials Science and Engineers*, Dallas, TX, 1998, pp. 78, 96.
34. M.I. Learada, *Revista de Plastics Modernos*, 2002, 83, 501.
35. G.D. Pasquale, A.D. La Roas, A. Recca, S. Di Calro, M.B. Baison, and S. Facchetti, *Journal of Materials Science*, 1997, 32, 3021.

36. S.S. Pesetskii, M.B. Kaplan, V.E. Starzhinskii, et al., *Vestsi Akad Nauk, BSSR, Ser. Fiz-Tekhn, Nauk*, 1986, 35.
37. N. Damyanov, P. Karalov, and S. Damyanov, *Trans. Plovdiv University Physics*, 1975, 13, 117.
38. O.F. Onda, L.V. Kutiina, A.O. Kogerman, and N.E. Kharitinych, *Ukrain Khim. Zh.*, 1978, 44, 526 (Russian ed.).
39. V.V. Korshak, I.A. Gribova, S. Yu Nekrasov, S.A. Pavlova, P.N. Gubkova, and L. Yu Avetisyan, *Vysokomol. Soedin., Ser. A*, 1978, 20, 271.
40. L.L. Chervyatova, G.I. Motryuk, and A.A. Kachem, *Surface Phenomena in Polymers*, Naukova Dumka, Kiev, 1927, p. 161 (in Russian).
41. V.P. Wong, X.H. Gao, R.M. Wang, H.G. Liu, C. Yang, and Y.R. Xiang, *Reactive and Functional Polymers*, 2008, 68, 1170.
42. Z. Liu, K. Chen, and D. Yan, *Polymer Testing*, 2004, 23, 323.
43. H. Ardhnananta, H. Ismail, and T. Takeichi, *Journal of Reinforced Plastics and Composites*, 2007, 26, 789.
44. K. Stoefller, P.G. Lafleur, and J. Denault, *Proceedings of the 64th Annual Conference (ANTEC 2006)*, Charlotte, NC, May 7–11, 2006, p. 263.
45. J. Bandyopadhyany, S.S. Ray, and M. Bousimen, *Macromolecular Chemistry and Physics*, 2007, 208, 1979.
46. M.M. Ramos-Tejada, M.M. Galdinilo-Gonzles, R. Perea, and J.D.G. Duran, *Journal of Rheology*, 2006, 50, 995.
47. P. Cebe and J. Runt, *Polymer*, 2004, 45, 1923.
48. G.I. Belockoneva, S.S. Demchenko, L.A. Red'ko, A.A. Kachan, and L.L. Cheryvatsova, *Plasticheskie Massy*, 1995, 57.
49. M.T. Bryk, *Polymerization on the Solid Surface of Inorganic Substances*, Nankova Dumka, Kiev, 1981, p. 288 (in Russian).
50. M.T. Beyk and T.E. Hipalova, *Physicochemistry of Multicomponent Polymer Systems*, Nankova Dumka, Kiev, 1986, pp. 9–82, 324 (in Russian).
51. M.T. Bryk and A.F. Bupban, *Polymer Composites*, Walter de Gruyter, Berlin, 1986, Vol. 11, 269.
52. M. Lai, J. Li, J. Yang, J. Liu, X. Tong, and H. Cheng, *Polymer International*, 2004, 53, 1479.
53. M.R. Loos, S.H. Pezzin, S.C. Amico, C.P. Bergmann, and L.A.F. Coelho, *Journal of Materials Science*, 2008, 43, 6064.
54. H.S. Kim, B.H. Park, J.S. Yoon, and H. Jin, *European Polymer Journal*, 2007, 43, 1729.
55. D. Wu, L. Wa, Y. Sun, and M. Zhang, *Journal of Polymer Science, Part B: Polymer Physics*, 2007, 45, 3137.
56. M.A. Magrupov, B.D. Yusupov, M.K. Akkarov, I. Gufurov, and U. Abdurakhumanov, *Usokomol Soedin, Ser. A*, 1976, 18, 2203.
57. M. Kalnin and V.P. Karlivan, *Suedin Sera*, 1968, 10, 2335.
58. Y. Malers and M.M. Kalmin, *Modification of Polymer Materials*, 1972, 3, 53 (in Russian).
59. M.D. Sizova, V.G. Rakova, M.N. Valiotti, V.I. Sergeev, N.N. Goraderskaya, and Y. Karmilova, *Vysokomol. Soedin., Ser. B*, 1986, 28, 406.
60. I.B. Kovarskaya and A.J. Sanz Larovsky, *Plasticheskie Massy*, 1971, 37.
61. Y. Malers and M.M. Kalinin, *Modifik. Polimer. Materialov* (Riga), 1980, 9, 5.
62. M.M. Kalnins, V.P. Karlivane, and I. Tiltinya, *Modfik Polimer Materials*, 1969, 2, 14.
63. M.M. Kalins, V.P. Karlivane, R. Brakere, and E. Metnica, *Vysokomol/Soedin., Ser. A*, 1967, 9, 2178.
64. M.M. Kalins, V.P. Karlivane, R. Brakere, and E. Metniece, *Vysokomol. Soedin., Ser. A*, 1967, 9, 2676.
65. R.A. Shanks, *Brti. Polym. J.* 1986, 18, 75.

66. G.I. Konovalova, D.N. Pacharzhanov, D.N. Kalontarov, I. Ya, and V.N. Naumov, *Plasticheskie Massy*, 1970, 42.

67. V.P. Plehanov, S.M. Berlyant, G.A. Burinkhinc, and V.Y. Vysokand, *1982 Sira* (982), 1982, 24, 1290.

68. M.A. Nagrupov, B.D. Yusupnt, and M.R. Atodzhanev, *Ush Khim Zhur*, 1971, 34.

69. V.P. Yarlsev, V.I. Istomin, and Yn V. Stretelkov, *Nauch. Trudy*, Moscow, In-la Mashinostr, 1975, 68, 76.

70. M.M. Volokhonskaya, V.B. Kopylov, and O. Sorokin, *Zhur Prikl., Khim.*, 1983, 56, 1198.

71. V.P. Yartsev, V.I. Istomin, and Yn V. Stretlkov, *Nauch. Trudy*, Moscow, In-la Mashinostr, 1975, 68, 78.

72. R.A. Lamb, *British Polymer Journal*, 1986, 18, 75.

73. G.I. Konovalova, D.N. Pacharzhanov, I. Ua Kalontarov, and V.N. Nanmov, *Plasticheskie Massy*, 1970, 42.

74. S.S. Pesetskii, M.P. Kaplan, and V.E. Starzhinskii, *Vesti Akad Nauk, BSSR, Ser Fiz-Tekhn-Nauk*, 1986, 15.

75. N. Damyonov, P. Karctalov, and S. Damyanov, *Trans. Plovdiv Univ. Physics*, 1975, 13, 117.

76. M.D. Sizova, V.R. Rakova, V.N. Valiotti, V.I. Sergeer, and M.N. Gorodeeskaya, *Vysokomol. Soedin., Series B*, 1986, 28, 408.

77. M.A. Magrupov, B.D. Yusunpov, and M.R. Atadzhanov, *Ush Khim zhur*, 1971, 34.

78. G.I. Belokoneva, S.S. Remchenko, L. Rediko, A.A. Kachan, and L.L. Chervatsova, *Plasticheskie Massy*, 1975, 57.

79. I.M. Abramova, L.O. Bunino, V. Vasil'eva, L.A. Zezina, I. Kazaryan, G. Kariniko, and V.I. Sergeer, *Plasticheskie Massy*, 1986, 8.

80. G. Ivan and M. Guirginica, *IUPAC Macro '83*, Bucharest, 1983, section 2-3, sp. 148.

81. G.P. Gladyshev, N.I. Mashukov, A.K. Mikitaev, and S.A. El'tsin, *Vysokomol. Soedin., Ser. B*, 1986, 28, 62.

82. H. Wagner, S. Sack, and E. Steger, *Acta Polymer*, 1982, 34, 65.

83. S. Sack, H. Wagner, and E. Steger, *Acta Polymer*, 1985, 36, 305.

84. C. Vasile, P. Omu, V. Barboin, and M. Sabliouschi, *Acta Polymer*, 1985, 36, 543.

85. I.B. Kovarskaya and A.T. Sanzharovsky, *Plasticheskie Massy*, 1971, 37.

86. Y.Y. Malers, and M.M. Kalinin, *Modifik. Polimer. Materialov* (Riga), 1980, 9, 5.

87. M.M. Kalnins, V. Karlivane, R. Brakere, and W. Metniece, *Vysokomol. Soedin., Ser. A*, 1967, 9, 2178.

88. A.A. Kachan, L.L. Chervyatsova, M.T. Bryk, and L.A. Red'ko, *Synthesis and Physicochemistry of Polymers*, 1974, 13, 131 (in Russian).

89. K. Murakami, S. Tamura, and T. Kusano, *Rep. Progr. Polym. Phys. Jpn.*, 1969, 12, 282.

90. K. Murakami and S. Tomura, *Kaoge Kagaki Dzassi.*, 1970, 73, 574.

91. K. Murakami, *Sikidzaj Kvekaisi*, 1972, 45.

92. A. Ballistreri, A. Foti, P. Margagvigna, G. Montaudv, and E.E. Scamporrino, *Journal of Polymer Science, Part B: Polymer Chemicals*, 1980, 18, 3101.

93. A. Dasgupta and S.K. Bhaltachrya, *London Journal of Technology*, 1982, 20, 68.

94. T. Kuila, M. Acharya, S.K. Srivastava, and A.K. Bhowmich, *Journal of Applied Polymer Science*, 2008, 108, 1329.

95. G.M. Butyrin, *High Molecular Weight Carbon Materials*, Khimiya, Moscow, 1976 (in Russian).

96. V. Vershovskiz, *Results of Science and Engineering Chemistry and Production of High Molecular Weight Compounds*, 1976, 67 (in Russian).

97. B.N. Smirnov, L. Tyan, A. Fialkov, T.U. Galkina, and G.S. Galeev, *Russian Chemistry Review*, 1976, 45, 884.

98. V.A. Mayatsky, L.B. Sokolov, and E.S. Soldatov, *Plasticheskie Massy*, 1982, 31.

99. R.F. Zinatullin, A.E. Egorov, Kh.F. Aminiev, and R.G. Nigmatullin, *Thesis of Reports of Conference of Young Scientists*, Ufa., 1986, p. 87 (in Russian).

100. V.V. Korshak, V.A. Sergeev, I.A. Gribova, S.A. Pavlova, I.V. Zhuravleve, T.A. Kolosova, et al., *Vysokomol. Soedin., Ser. A*, 1980, 22, 1228.

101. G.P. Gladyshev, *Methods for Stabilisation of Thermally Resistant Polymers*, preprint M, 1972 (in Russian).

102. E.A. Goldovsky and A.A. Dontsov, *Kauchuk I Rezina.*, 1980, 42.

103. R.F. Willis and R.F. Shaw, *Journal of Colloid Interface Science*, 1969, 31, 379.

104. V.P. Yartgov, V.I. Vistomin, and Yu V. Stretelkov, *Nauch Trudy,* Moscow, In te Mashinostr, 1975, 68, 76.

105. R. Salovey and W.A. Yager, *Journal of Polymer Science, Part A: General Papers*, 1964, 2, 219.

106. Y. Hajimoto, N. Tamura, and S. Okamoto, *Journal of Polymer Science, Part A: General Papers*, 1965, 3, 255.

107. I. Ouchi, *Journal of Polymer Science, Part A: General Papers*, 1965, 3, 2685.

108. Y. Sakai and M. Iwasaki, *Journal of Polymer Science, Part A: Polymer Chemistry*, 1969, 7, 1749.

109. Y. Hama and K. Shinohara, *Journal of Polymer Science, Part B: Polymer Physics*, 1970, 8, 651.

110. G. Gallet, S. Carroccio, P. Rizzarelli, and S. Karlsson, *Polymer*, 2002, 43, 1081.

7 Applications of Reinforced Plastics

7.1 SUMMARY OF APPLICATIONS

In Table 7.1, the summary of applications of reinforced plastics are reviewed under three main headings: mechanical, thermal, and electrical. Reinforcing agents considered include glass fibers, carbon fibers, carbon nanotubes, graphite, talc, minerals, silica, mica, and calcium carbonate. These applications now cover a wide range of applications, including general engineering, automotive, aerospace, building materials, electronics and microelectronics, power sources, medical, and bioengineering.

Some more recent developments in the uses of nanocomposites are reviewed in Table 7.2.

A rapidly developing area is the use of nanocomposites in the manufacture of sports equipment.

One of the initial uses for exfoliated clay in barrier applications involved a 20 μm coating on the interior of a tennis ball to prevent depressurization. The product was developed by InMat LLC and introduced in 2001. Sports equipment was one of the initial areas where carbon fiber composites were commercialized. This is also true for carbon nanotubes, where the carbon nanotubes (at low levels) reinforce the epoxy matrix of the carbon fiber composite in speciality tennis rackets and hockey sticks. In such applications, performance overrides the economic disadvantages of the expensive carbon nanotube inclusion.

7.2 APPLICATIONS USING GLASS FIBER–REINFORCED POLYMERS

More detailed information on the application of glass fiber reinforcement is given in Table 7.3. It can be seen that glass fiber concentrations used are from as low as 10% in polyethylene oxide automotive improvements, panels, and polyetherimide carbon fans and gears to as high as 50%–60% in polyethylene oxide gears, sprockets, propellers, and automotive bumpers. Polyamide 6,6 is used in high-strength and -stiffness applications.

Many of these applications require polymers meeting particularly high property standards, for example, stability, tensile strength and impact properties, thermal properties, dimensional stability, and chemical and oil resistance.

7.2.1 POLYAMIDES

An appreciable amount of work has been carried out on the use of polyamides in engineering applications such as automotive parts, precision engineering applications, and high-strength and -stiffness applications such as fan blades, gears, and bearings.

TABLE 7.1
Some Applications of Reinforced Plastics

Polymer	Reinforcing Agent	Mechanical	Electrical	Thermal	Other
Polypropylene	20% talc	Stiffness at elevated temperature, automotive under-bonnet application, air ducting channels	Electrical systems housing		Housing from kettles
Polypropylene	20% glass fibers	Automotive under-bonnet application	—	Cooling system expansion tanks	—
Polypropylene	20% glass fibers and 40% $CaCO_3$				Garden furniture
Polypropylene	20% glass fiber, short glass fiber reinforced	—	—	—	Washing machine parts
Alkyd resin	Short glass fiber reinforced	Switching relay boxes	Electronic switches, automotive ignition, switch relay boxes		
Alkyd resin	Short glass fiber reinforced	Switching insulators	Electrical switching	—	—
Alkyd resin	Mineral and fiber reinforced	Relay capacitive	Switches, relay bases, capacitive insulation	Microwave oven	
Alkyd resin	Short glass fiber	Automotive ignition electronic switches, switch relay base	—	—	—

Polymer	Filler	Mechanical	Electrical	Thermal	Miscellaneous
Polyethylene terephthalate	45% mineral glass fiber reinforced	Automotive ignition	Electrical components, terminal blocks	—	Function for furniture components
Polyethylene terephthalate	30% glass fiber reinforced Housing for pumps, motors, etc., gears, sprockets	Electrical components	—	—	
Polyethylene terephthalate	30% glass fiber reinforced	Exterior	automotive body parts, casing and housing, water pump housings, windscreen wipers, wiper systems	Electrical component	
Polybutylene terephthalate	10%–30% glass fiber reinforced	Automotive distribution caps	Lamp socket switches, motor housing	Telecommunications equipment	—
Polybutylene terephthalate	30% glass fiber reinforced	—	Electrical motor parts, switches	—	—
Polybutylene terephthalate	30% glass bead reinforced	Engineering parts	—	Heating appliances	—
Polybutylene terephthalate	45% and mineral glass fiber reinforced	—	—	Oven grills	—
Polybutylene terephthalate	30% carbon fiber reinforced	Ultra-high-strength and -stiffness applications	—	—	—
Polyether ether ketone	20% glass fiber	Automotive application	Electrical components, printed circuit boards	High temperature application	Aerospace
Polyether ether ketone	30% glass fiber	Pump housing, pipe fittings	Electrical components, printed circuit boards	—	Missile aircraft nose cones
Polyether ether ketone	30% mica and mica and glass fiber reinforced	High-dimensional stability components, structural housing	Electrical components	—	
Polyether ether	30% carbon fiber application	Automotive engines, pump impellers, bearings, bushes	—	—	Structural aerospace

Continued

TABLE 7.1 (Continued)
Some Applications of Reinforced Plastics

Polymer	Filler	Mechanical	Electrical	Thermal	Miscellaneous
Polyoxymethylene	Glass fiber reinforced	Pump housing, pipe fittings	—	—	Sporting equipment
Polyoxymethylene	PTFE lubricated	Bearings, fan, blades	Switch components	—	—
Polyoxymethylene	30% glass fiber reinforced	Gears, bearings, brushes	—	—	Sporting equipment
Polyoxymethylene	30% carbon fiber reinforced	Bearings, bushes, gears	—	—	—
Polyetherimide	30% glass fiber reinforced	Explosion-proof containers, automotive underbody applications, heat exchangers, fuel systems	Electrical switches	Thermal protectors	—
Polyamide imide	Glass fiber reinforced	Valve plates, pistons, rotors	Terminal strips, insulations	—	—
Polyamide imide	Graphite filler	Bearings, washers, piston rings, seals, impellers	—	—	—
Polybutylene sulfide	30% carbon fiber reinforced	Pump housings, valves, high-stress structures	—	High heat applications	Chemical resistance
Polyphenylene sulfide	Glass fiber and head reinforced	Automotive head lamps; under bonnet; parts exposed to oil, petrol, and hydraulic fluids; pumps; valves; precision mechanical parts	Boxes of electronic circuits, relays, breakers	High heat applications	High chemical resistance
Polyphenylene sulfide	40% glass fiber reinforced	Automotive head lamps, petrol, oil, hydraulic fluid–resistant parts, exhaust gas, emission control valves, pumps, precision mechanical parts	Terminal blocks, connectors, electric motor housing boxes for electronic components	High heat application	High chemical resistance

Material	Grade	Application	Electrical	Heat-resistant	
Polyamide 6,6	MoS₂ lubricated	Bearings, cams	Electrical parts	—	—
Polyamide 6	30% glass fiber reinforced	Automotive parts	—	—	—
Polyamide 6	30% carbon fiber reinforced	Bearings, high-strength structure	—	—	—
Polyamide 6/6	40% mineral filled	Under-bonnet mechanical parts	—	—	—
Polyamide 6/6	60% glass fiber reinforced	High-strength and -stiffness applications	—	—	—
Polyamide 6/6	MoS₂ lubricated	Bearings, cams, valves, valve seals, gears	—	—	—
Polyamide 6/6	Carbon fiber reinforced	Connecting rods	—	—	—
Polyamide 6/10	3% glass fiber reinforced	Precision gears, cams, bearings, valve seals	Electrical connection couplings	—	—
Polyamide 6/10	Carbon reinforced	Precision engineering parts	Electrical components	—	—
Polyamide 6/11	30% glass fiber	Fan blades, gears, precision engineering parts, hoses	Electrical plugs	Heat-resistant parts	—
Polyamide 6/12	20% PTFE lubricating	Gears, mechanical parts, low water absorption	—	—	—
Polyamide 6/12	10%–30% glass fiber reinforced	Mechanical parts, low motor absorption	—	—	—
Polyamide 12	30% glass fiber reinforced	High precision and stability applications, fan blades	Electrical plugs	Heat-resistant housings	—
Polyamide 12	50%–60% glass bead reinforced	Rigid dimensional stability parts, bushes, valves, etc.	Electrical coils and bobbins	—	—
Polyamide	MoS₁ lubricated	Machine bearings, insulating bushes, drive rollers	—	—	—
Polyimide	25% graphite reinforced	Valve seals, compressor rings reinforced	Gear boxes	Terminal breakers	—
Polyimide	40% glass fiber				—

Continued

TABLE 7.1 (Continued)
Some Applications of Reinforced Plastics

Polymer	Filler	Mechanical	Electrical	Thermal	Miscellaneous
Polyetherimide	10% or 20% glass fiber reinforced	Cooling fans, gears, automotive under-body components, fuel systems	Electrical components, fuse	High-temperature connections, heat exchangers	—
Fluorinated ethylene-propylene	20% glass fiber reinforced	Valve components	Electrical	—	Cleaning plant
Silicones	Glass fiber and mineral filled	—	Electronic component encapsulations	—	—
Polytetrafluoroethylene	60% bronze filled	High-speed bearings, wear pads, piston rings	—	—	—
Polytetrafluoroethylene	15% or 25% glass fiber reinforced	Wear pads, piston rings	—	—	Microwave oven parts
Polytetrafluoroethylene	15% graphite filled	Bearings, wear pads, piston rings	—	Antistatic applications	—
Ethylene chlorotrifluoroethylene	Glass fiber reinforced	Valves	—	—	Chemical containers
Polytetrafluoroethylene	10% + 30% glass fiber reinforced	Bearings in aggressive environment pumps, impellers, valves, gears	—	—	—
Polytetraethylene fluoroethylene	30% carbon fiber filled	Bearings, seals, rings, piston rings, valve plugs, compressor rings	—	—	—

Material	Reinforcement	Application	Electrical application	Other
Perfluoroalkoxy ethylene	20% glass fiber reinforced	—	—	Chemical plant Use in aggressive environment
Polyvinylidene fluoride	20% carbon fiber reinforced	Bearings, nuts, pump rotors, valve bodies	—	—
Polyphenylene sulfide	20% PTFE lubricated	Antifriction gears, bolts, and screws	—	—
Polyethersulfone	30% glass fiber reinforced	Under-bonnet automotive applications, fittings and connectors, in gear box, area air ducting, car heating fans	Electrical components, printed circuit boards	—
Polysulfone	30% glass fiber reinforced	Pumps, valves in petrochemical industry	Electrical components, circuit boards	—
Polysulfone	10% glass fiber reinforced	Automotive under-body application	Electrical components, printed circuit boards, electronic ignition components	Aerospace application, alkaline battery cases
Polysulfone	30% carbon fiber reinforced	Under-bonnet automotive, chemical plant, pumps, valves, process pipes	Switch devices	Circuit interior and exterior components
Polysulfone	15% PTFE lubricated	Pumps, valves	Electrical components, terminal blocks, housing fan, electrical components	Medical equipment heading sterilization

TABLE 7.2

Polymer Matrix	Nanoparticle	Property Improvement	Application	Company or Product Trade Name
Polyamide 6	Exfoliated clay	Stiffness	Timing belt cover: automotive	Toyota/Ube
TPO (thermoplastic polyolefin)	Exfoliated clay	Stiffness/strength	Exterior step assist	General Motors
Epoxy	Carbon nanotubes	Strength/stiffness	Tennis rackets	Babolat
Epoxy	Carbon nanotubes	Strength/stiffness	Hockey sticks	Montreal: Nitro Hybonite®
Polyisobutylene	Exfoliated clay	Permeability barrier	Tennis balls, tires, soccer balls	InMat LLC
Styrene butadiene resin (SBR), natural rubber, polybutadiene	Carbon black (20–100 nm: primary particles)	Strength, wear and abrasion	Tires	Various
Various	MWCNT	Electrical conductivity	Electrostatic dissipation	Hyperion
Unknown	Silver	Antimicrobial	Wound care/ bandage	Curad®
Nylon MXD6, PP	Exfoliated clay	Barrier	Beverage containers, film	Imperm™: Nanocor
SBR rubber	Not disclosed	Improved tire performance in winter	Winter tires	Pirelli
Natural rubber	Silver	Antimicrobial	Latex gloves	
Various	Silica	Viscosity control,	thixotropic agent	Various
Polyamide nylons 6, 12, 66	Exfoliated clay	Barrier	Auto fuel systems	Ube

Several workers have discussed the mechanical properties of reinforced plastics [2–8]. Keating et al. [3] checked the necessity of high-temperature annealing and the degree of its impact on creep straining glass fiber–reinforced polymers polyamide 6,6 and polyethylene terephthalate, which had glass transition temperatures of 58°C and 132°C, respectively.

The improvement after annealing depends on the polymer type and whether it is fast or slow crystallizing at the molding process and whether it has hydrogen-bonded sheets. The physical property changes observed of these materials before and after annealing support the explanation of crystal reorganization through crystallization, free volume reduction through densification, and crystal perfection through better chain packing. These data were used to predict long-term creep strains up to 10 years

TABLE 7.3

Glass Fiber– or Glass Bead–Reinforced Plastics

Polymer	Reinforcing Agent	% Reinforcing Agent	Application Agent
Polypropylene	Glass fiber	20	Automotive under-bonnet application
Polyphenylene oxide	Glass fiber	10	Automotive instrument panels
Epoxy resins	Glass filled	—	Power transmission equipment
Epoxy resins	Glass prepag	—	Building products, pressure vessels
Polyether ether ketone	Glass fiber	20	Automotive application
Polyether ether ketone	Glass fiber	35	High-stability components, structural housings
Polyoxymethylene (polyacetal)	Glass fiber	30	Gears, bearings, bushes
Diallyl isophthalate	Long glass fiber	—	Pump impellers
Polycarbonate	Glass fiber	20	Replacement for metal parts
Polyethylene terephthalate	Glass fiber	50	Gears, sprockets, propellers, automotive bumpers
Polyethylene terephthalate	Glass fiber and mineral	45	Automotive ignition
Polyethylene terephthalate	Glass fiber	30–35	Pump housings, gears, sprockets, automotive exterior casing, housing for pumps
Polybutylene terephthalate	Glass fiber	10–20	Automotive distributor caps
Polyamide 6	Glass fiber	30	Automotive parts
Polyamide 6,6	Glass fiber	60	High-strength and -stiffness applications
Polyamide 6,10	Glass fiber	30	Precision engineering, gears, cams, bearings, valve seals
Polyamide 1,1	Glass fiber	30	Mechanical parts, fan blades, gears
Polyamide 6,12	Glass fiber	10–30	Mechanical parts, low-water-absorption applications, high-strength applications
Polyamide 12	Glass fiber	30	High-precision applications, fan blades
Polyamide 12	Glass fiber	50	Rigid dimensional stability, bushes, valves
Polyimide	Glass fiber	40	Gear boxes
Polyetherimide	Glass fiber	10	Cooling fans, gears, automotive under-body applications, fuel systems
Polyetherimide	Glass fiber	30	Explosion-proof containers, automotive under-body applications, heat exchangers, fuel systems
Polyamide-imide	Glass fiber	—	Valve plates, pistons, gears, rotors

Continued

TABLE 7.3
Glass Fiber– or Glass Bead–Reinforced Plastics

Polymer	Reinforcing Agent	% Reinforcing Agent	Application Agent
Polyethersulfone	Glass fiber	30	Automotive under-body applications, gear box parts, air ducting, car heater fans
Polysulfone	Glass fiber	30	Pumps, valves, in petrochemical industry
Polysulfone	Glass fiber	10	Automotive under-bonnet applications
Polytetrafluoroethylene	Glass fiber	15–25	Wear pads, piston rings
Ethylene-tetrafluoroethylene	Glass fiber	10–30	Bearings, valves, gears
Perfluoroalkoxy ethylene	Glass fiber	20	Pipes, bearings
Fluorinated ethylene-propylene	Glass fiber	20	Valves

using a time–temperature superposition technique. The accuracy of the prediction was shown with a confidence level of about 90%.

Karrajannidis et al. [2] and Khari and Kisman [9] studied crack propagation and impact strength of glass fiber and reinforced polyamide 6,6 containing up to 12.5% of a functionalized T-block copolymer based on styrene(ethylene-co-butylene)styrene grafted with maleic anhydride. Blends containing 2.5, 5, 7.5, 10, and 12.5 wt% copolymer were prepared by melt blending in a single-screw extruder. Emphasis was placed on mechanical properties in comparison with morphology and thermal properties of the aforementioned samples. Although the amount of styrene(ethylene-co-butylene)styrene grafted with maleic anhydride added to polymer 6,6 is not enough to produce a supertough material, a significant increase in the resistance to crack propagation and impact strength was observed in all blends. This behavior was proportional to the amount of styrene(ethylene-co-butylene)styrene–grafted maleic anhydride added for samples having up to 10%, rubber, while additional amounts seem to have no further effect. A small decrease in tensile strength was observed from spectroscopy, and it is shown that the grafting extent of styrene(ethylene-co-butylene)styrene maleic anhydride polyamide 6,6 is very low.

Seldon [10] measured the weld line strength and impact, and flexural and tensile properties of injection-molded specimens of glass fiber–reinforced polyamide 6 and talc-filled polypropylene (PP). Further information was obtained by examination of fracture surfaces by scanning electron microscopy.

Toensmeir [11] of Flightcoms Corp. claims that the terminal headset is the first aviation headset to make almost total use of plastics. Key properties required for the headset include low weight, high impact strength, resistance to extreme heat and cold, and a repeatable, comfortable head-clamping force. The design and development firm SoMa used finite element analysis (FEA) and design of experiments (DOE) methods to analyze cross sections and materials. Glass-reinforced polyamide was eventually specified for the headband. The domes or earpieces are molded acrylonitrile-butadiene-styrene, and the microphone housing and elbow are molded of acetal. The earpiece and headset cushions are polyurethane foam covered by vinyl cloth.

Gotzmann et al. [12] of Bayer Plastics discussed the manufacture of metal-plastic hydride components, including injection molding plastics such as glass fiber–reinforced polyamides around a steel aluminum profile placed in the mold. Some applications of such components were examined, and computer-aided material selection, finite element analysis, and computer simulation systems developed for use in this technology are described.

7.2.2 POLYKETONES

LNP Engineering Plastics [13] has introduced a series of formulations of Carilon-based thermoplastic compounds. LNP's new line will form part of the company's N-series, which contains glass and carbon fiber reinforcements and polytetrafluoroethylene as a lubricant. Carilon aliphatic polyketones are characterized by good impact performance over a broad temperature range, high chemical resistance, good hydrolytic stability, and superior resilience. Target markets include automotive, business machines, domestic appliances, and electrical applications.

7.2.3 POLYPHENYLENE SULFIDE

Ford [14] have developed an automotive fuel rail based on a linear partially crystalline 40% glass fiber–reinforced polyphenylene sulfide. This engine component conveys fuel from the injection pump to the cylinders and is required to withstand rapid changes of pressure, service temperatures of up to 120°C, engine vibrations, and direct contact with aggressive fuels. Details are given of chemical resistance tests carried out on the Fortron grade and also on the branched polyphenylene sulfide grade and two polyamide 6,6 grades. Test fuels comprised ASTM reference fuel C, two aggressive fuels with differing methanol proportions, and an auto-oxidized fuel containing peroxide and copper additives. Fortron 1140L4 exhibited the highest impact resistance and dimensional stability and is now used to produce fuel rails with reduced weight and cost.

Other polymers with a fuel resistance to solvents include 30% carbon fiber–reinforced polyphenylene sulfide, glass fiber–reinforced ethylene chlorotrifluoroethylene, 20% glass fiber–reinforced perfluoroalkoxy ethylene, 20% carbon fiber–reinforced polyvinylidene, and 20% glass fiber–reinforced fluorinated ethylene-propylene.

7.3 APPLICATIONS OF CARBON FIBER

As shown in Table 7.4, carbon fibers have been used to reinforce a wider variety of polymers used in a range of applications.

TABLE 7.4
Uses of 30% Carbon Fiber–Reinforced Plastics

Polymer	Applications
Polyether ether ketone	Bearings, cams, gears
Polyoxyethylene polyacetal	Bearings, gears
Polycarbonate	High-strength plastics
Polybutylene terephthalate	Ultra-high-strength and -stiffness applications
Polyamide 6	Bearings, high-strength structures
Polyamide 6,6	Connecting rods
Polyamide 6,10	Precision engineering parts, housing, valves
Polyamide 6,12	Precision engineering parts
Polyphenylene sulfide	High-strength structured components
Polysulfone	Automotive underbody applications, chemical plant, pumps, valves, tires, pump impellers
Ethylene tetrafluoroethylene	Bearings, seals, rings, piston rings, valves, plug compression rings
Polyvinylidene fluoride	Bearings, nuts, pump rotors
Polytetrafluoroethylene	Valve bodies
Epoxy resins	Structural components, boat hulls, chemical engineering plant, pressure vessels, aerospace structural components, helicopter blades, racing cars
Polyethersulfone	Aerospace application, nose cones and y-ducts, pump impellers

7.3.1 POLYAMIDES

Carbon fiber– and glass fiber–reinforced polyamides 6 have been used in the construction of wind turbine blades [15] using thermoset composite technology. It is claimed that such structures are stronger than thermostats for the same weight, and so permit lighter structures.

7.3.2 POLYETHER ETHER KETONE

LNP Engineering Plastics [13] have introduced a series of formulations based on polyether ether ketone reinforced with glass and carbon fibers using polytetrafluoroethylene (PTFE) as a lubricant. These polymers have very good impact performance over a wide range of temperatures and have high chemical resistance, good hydraulic stability, and superior resilience. These polymers are targeted at the automotive, business machine, domestic appliances and electronics markets.

A grade polyether ether ketone, developed for medical implant applications and marketed by Invibio as PEEK-OPTIMA, has over recent years [16] found increasing application as a neat polymer and as a product containing carbon fibers. Extensive testing has demonstrated the biocompatibility of these materials in vitro and in vivo to the satisfaction of U.S. and European notified bodies. Increasing numbers of products are now CE marked and Food and Drug Administration (FDA) approved. Further supporting evidence is provided on PEEK-OPTIMA's biocompatibility, but also biostability and resistance to gamma irradiation and long-term oxidation. Chemical analysis by Fourier transform infrared spectroscopy (FTIR) and Gel permeation chromatography (GPC) of explanted material (aged 10 years and implanted in rabbit paravertebral muscle for 1 year) shows that the material is biostable, with no new chemical species present, and that the molecular weight distribution does not change after implantation. Gamma sterilization does not affect the mechanical properties of the polymer over a wide range of standard tests, and oxidation testing (simulating 10-year aging) demonstrates that this material is very resistant to oxidation, unlike ultra-high-molecular-weight polyethylene. The overall conclusion is that PEEK-OPTIMA is a very stable polymer that is well suited for use in human implantable devices.

7.3.3 PHENOLIC RESINS

Patton et al. [17] evaluated the ablation, mechanical, and thermal properties of vapor-grown carbon fiber phenolic resin (Pyrograf III from Applied Sciences Inc./ SC-1008). These composites were evaluated to determine the potential of using this material in solid rocket motor nozzles. Composite specimens with varying vapor-grown carbon fiber loadings (30–50 wt%), including one sample with ex-rayon carbon fiber plies, were prepared and exposed to a plasma torch for 20 s with a heat flux of 16.5 MW/m^2 at about 1650°C. Low erosion rates and little char formation were observed, confirming that these materials were promising for use as rocket motor nozzle materials. When fiber loadings increased, mechanical properties and ablative properties improved. The vapor-grown carbon fiber composites had low thermal conductivities (about 0.56 W/m-K), indicating that they were good insulating materials. If a 65% fiber loading in vapor-grown carbon fiber composites could be achieved, then ablative properties were expected to be comparable to or better than the composite material currently used in the Space Shuttle Reusable Solid Rocket Motor.

7.4 APPLICATIONS OF CARBON NANOTUBES

Some applications of carbon nanotube–reinforced plastics are tabulated in Table 7.5.

In addition to the mechanical properties, carbon nanotube–reinforced polymers have interesting electrical properties, in particular, their electrical conductivity (see Section 5.8).

Epoxy resins have wide engineering applications due to their low cost; easy processability; good thermal, mechanical, and electrical properties; and so forth. With the rapid developments in spacecraft and superconducting cable technologies, and large cryogenic engineering projects such as the International Thermonuclear Experimental Reactor, epoxy resins have been increasingly employed in cryogenic engineering applications as impregnating materials, adhesives, and matrices for advanced composites [26–33]. However, pure epoxy resins normally have poor crack resistance at room temperature [34] and could be more brittle at cryogenic temperatures [35–38], which makes them unsuitable for some cryogenic engineering applications that demand epoxy resins to have high cryogenic mechanical properties. For example, during service of epoxy resins at an International Thermonuclear Experimental Reactor, the temperature might change from room temperature to cryogenic temperatures (like liquid nitrogen temperature, 77 K, etc.) or vice versa

TABLE 7.5
Electrical Properties of Carbon Nanotube/Polymer–Based Polymers

Polymer	Property Measured	Reference
Polyethylene carbon nanotube	Electrical conductivity and rheological properties	18
Epoxy resin	Mechanical and electrical properties	19
Poly(p-phenylene vinylene)	Electrical and optical properties	20
Polylactide composites	Electrical conductivity	21
Polyurethane nanocomposites	Electrical conductivity surface resistivity	22
Nanocomposites	Mechanical properties	—
Polyether ether ketone	Rheological properties and electrical conductivity	23
Polyetherimide and epoxy resins	Mechanical properties and electrical conductivity	24
Polycarbonate	Rheological and electrical properties	25

[39–41]; this would induce thermal stresses within epoxy resins and demand epoxy resins to have high cryogenic mechanical properties to withstand internal thermal stresses. Therefore, it is of great importance to improve the cryogenic mechanical properties of epoxy resins so that they can be gainfully used in cryogenic engineering applications. Recent publications have reported on epoxy toughening and strengthening of epoxy resins for enhancing cryogenic mechanical properties using flexible diamine thermoplastic polyethersulfone [38], silica nanoparticles [42], exfoliated montmorillonite [43], hyperbranched polymer [44], polyurethane [45], and n-butyl glycidyl ether [46].

Carbon nanotubes are long cylinders of covalently bonded carbon atoms and have a diameter from a few angstroms to several tens of nanometers. Carbon nanotubes have exceptional mechanical properties [47–50], and extensive research work has been carried out on carbon nanotube–reinforced polymer composites [46–52]. However, weak interfacial bonding between carbon nanotubes and polymers leads to poor stress transfer, and this has limited the full realization of carbon nanotubes as reinforcements for polymers. Therefore, chemical functionalization of carbon nanotubes has been conducted.

Chen et al. [55] have applied epoxy resins in cyrogenic engineering. It was shown that cryogenic properties can be improved by the incorporation of carbon nanotubes in the polymer formulation. Carbon nanotubes are regarded as exceptional reinforcements for polymers. However, poor carbon nanotube–polymer interfacial bonding leads to an unexpected low reinforcing efficiency. Chen et al. [55] studied the cryogenic mechanical properties of multiwalled carbon nanotube (MWCNT)–reinforced epoxy nanocomposites, which are prepared by adding multiwalled carbon nanotubes to diglycidyl ether of bisphenol F epoxy via the ultrasonic technique. When the temperature decreased from room temperature to liquid nitrogen temperature (77 K), a strong carbon nanotube–epoxy interfacial bonding was observed due to the thermal contraction of the epoxy matrix because of the big differences in thermal expansion coefficients of epoxy and multiwalled carbon nanotubes in the brittle primary phase, but not in the soft second phase in the two-phase epoxy matrix. Consequently, the cryogenic tensile strength, Young's modulus, failure strain, and impact strength at 77 K are all enhanced by the addition of multiwalled carbon nanotubes at appropriate contents. The results suggest that carbon nanotubes are promising reinforcements for enhancing the cryogenic mechanical properties of epoxy resins that have potential applications in cyrogenic engineering.

Kim et al. [56] also discussed the cryogenic properties at −150°C of carbon black nanoparticle–reinforced epoxy resins. Nanoparticle reinforcement improved fracture toughness at room temperature but decreased fracture toughness at cryogenic temperatures.

7.4.1 SULFONATED (POLYARYLENE SULFONE)

Joo et al. [53] have discussed a new type of composite membrane, consisting of functionalized carbon nanotubes (CNTs) and sulfonated polyarylene sulfone (SPAS) for the construction of a direct methanol fuel cell. The carbon nanotubes modified with sulfonic acid or platinum-ruthenium in nanoparticles were dispersed with the sulfonated polyarylene sulfone matrix by a solution-casting method to afford SO″ 3CNT-SPAS or Pt Ru/CNT-SPAS composite membranes, respectively. Characterization of the composite membrane reveals that the functionalized carbon nanotubes are homogeneously distributed within the sulfonated polyarylene sulfone matrix and the composite membranes contain smaller ion clusters than the neat sulfonated polyarylene sulfone. The composite membrane exhibits enhanced mechanical properties in terms of tensile strength, strain, and toughness, which leads to improvements in ion conductivity and methanol permeability compared with the neat sulfonated polyarylene sulfone membrane. In a direct methanol first-cell performance test, the use of a carbon nanotube-sulfonated polyarylene sulfone membrane yields higher power density.

Compared to the polymer sulfone membrane, which demonstrates that the improved properties of the composite membranes induce an increase in power density, the strategy for carbon nanotube-sulfonated polyarylene sulfone composite membranes discussed here could potentially be extended to other carbon nanotube–polymer composite systems. Recent advances in the chemistry of carbon nanotube functionalization [54], and further understanding of the interaction between the carbon nanotubes and the polymer, may encourage the preparation of a variety of new composite materials.

Carbon nanotubes are being increasingly used for the reinforcement of plastics, including epoxy resins [57–59], polyamides [55, 60–62], polyimides [63], polysilsesquioxane [64], polycaprolactam [65], polyethylene oxide [66], polyethylene [67, 55, 62], polyurethane [67], ethylene-vinyl acetate [69–73], polyhydroxybutylene-co-hydroxyvalerate [75], ethylene-propylene diene terpolymer [73], and sulfonated polyarylene sulfone [53].

7.5 ELECTRONICALLY CONDUCTING POLYMERS

Guimard et al. [74] have carried out a very detailed review on developments in electrically conducting polymers, particularly in the field of biomedical engineering.

Conducting polymers have electrical and optical properties similar to those of metals and inorganic semiconductors, but they also exhibit the attractive properties associated with conventional polymers, such as ease of synthesis and processing. This unique combination of properties has given these polymers a wide range of applications in the microelectronics industry, including battery technology, photovoltaic devices, light-emitting diodes, and electrochromic displays, and more recently in the biological field. Research on conducting polymers for biomedical applications expanded greatly with the discovery in the 1980s that these materials were compatible with many biological molecules, such as those used in biosensors. By the mid-1990s, conducting polymers were also shown, via electrical stimulation, to modulate cellular activities, including cell adhesion, migration, DNA synthesis, and protein secretion. Specifically, many of these studies involved nerve, bone, muscle, and cardiac cells, which respond to electrical impulses. Most conducting polymers present a number of important advantages for biomedical applications, including biocompatibility, ability to entrap and controllably release biological molecules (i.e., reversible doping), ability to transfer charge from a biochemical reaction, and the potential to easily alter the electrical, chemical, physical, and other properties of the conducting polymers to better suit the nature of the specific application. These unique characteristics are useful in many biomedical applications, such as biosensors, tissue engineering scaffolds, neural probes, drug delivery devices, and bioactuators.

Polymers that, to date, have been investigated for their electrical conducting properties include polypyrrole, polyaniline, polythiophene, and polymer nanotube composites. The advantages of conducting polymers include good conductivity, biocompatibility, good stability, low impedance ability to entrap molecules, efficient charge transfer, and ability to entrap biomolecules.

REFERENCES

1. LNP Engineering, *Modern Plastics International*, 2001, 31, 65.
2. G. Kareyannis, D.N. Bikiaris, and G.Z. Papageorgiou, *Modern Plastic International*, 2002, *202*, 101.
3. M.Y. Keating, U.B. Malone, and W.D. Saunders, *Journal of Technical Analysis and Calorimetry*, 2002, 69, 6937.
4. A. Licsea-Claverie, F.J.U. Larrillo, A. Avarez-Castillo, and V.M. Castano, *Polymer Composites*, 1999, 20, 314.
5. E. Klata, S. Barysiak, K. Vandervelde, J. Garbarczyk, and J.K. Krucinska, *Fibres and Textiles in Eastern Europe*, 2004, 12, 64.
6. R. Seldon, *Polymer Engineering and Science*, 1997, 37, 205.
7. G. Gotzmann, *Revenue Generale des Caoutchoucs et Plastiques*, 1996, 148, 37.
8. P.A. Toensmeir, *Modern Plastics International*, 2001, 31, 108.
9. B. Khari, A. Kusman, M. Ishak, W.J. Chan, and T. Taleishi, *European Polymer Journal*, 2008, 44, 102.
10. R. Seldon, *Polymer Engineering and Science*, 1997, 37, 205.
11. P.A. Toensmeir, *Modern Plastics International*, 2001, 31, 108.
12. G. Gotzmann, *Revenue Generale des Caoutchoucs et Plastiques*, 1996, 148, 37.
13. LNP Engineering Plastics, *Plastics and Rubbers Weekly*, August 8, 1997, p. 13.
14. F. Ford, Walbro Corp, Ticon, J.M.B.H. Heochst A.G.
15. *Reinforced Plastics Journal*, February 2010, p. 31.

16. G. Green and K. Cartwright, *Proceedings of the Medical Polymer Conference*, Dublin, April 2–3, 2003, paper 5, p. 33.
17. R.D. Patton, C. Pitman, L. Wong, J.R. Hall, and A. Dams, *Composites, Part A: Applied Science and Manufacturing*, 2002, 34A, 243.
18. X. Lian-Chen, H.-Y. Fu, L. Zhang, and S.-Y. Zhu, *Polymer Materials Science and Engineering*, 2008, 24, 87.
19. S.-M. Yuen, C.-C. Ma, C.-Y. Chung, Y.H. Hsiao, C.-L. Chiang, and A.-D. Yum *Composites, Part A*, 2008, 39, 1.
20. H. Aarab, M. Baitoul, J. Wery, R. Aimairac, S. Hefrant, E. Faulques, J.L. Duval, and M. Hamedoun, *Synthetic Metals*, 2005, 155, 63.
21. C.F. Kuan, C.H. Chen, H.C. Kuan, and K.C. Lin, *Proceedings of 65th SPE Annual Conference (ANTEC 2007)*, Cincinnati, OH, May 6–11, 2007, p. 2250.
22. Hsu-Chiang Kuan, Chen Chi, Ma Composites technologies for 2020 Procedures of Fourth Asian-Australian Conference on Composite Materials (AA CMU) held Sydney Australia 6–9th July, Cambridge Woodhead Publishers 2004, p. 736.
23. D.S. Banarusanmpath, V. Altstaedt, H. Ruckdoeschel, J.K.W. Sandler, and M.S.P. Shaffer, *Journal of Plastics Technology*, 2008, 6, 19.
24. T. Aoki, Y. Ono, and T. Ogasawara, *Proceedings of the American Society for Composites, Twenty First Technical Conference*, Dearborn, MI, September 17–20, 2006, paper 31.
25. Y.T. Sung, M.S. Han, K.H. Song, J.W. Jung, H.S. Lee, C.K. Kim, J. Joo, and W.N. Kim, *Polymer*, 2006, 47, 201.
26. P.E. Fabian, J.B. Shutz, C.S. Hazleton, and R.P. Reed, *Advances in Cryogenic Engineering*, 1994, 40, 1007.
27. D. Evans and S.J. Canger, *Advances in Cryogenic Engineering*, 2000, 46, 361.
28. Q. Chen, B.J. Gao, and J.L. Chen, *Journal of Applied Polymer Science*, 2003, 89, 1385.
29. Y. Shindo, S. Takano, K. Horiguchi, and T. Sato, *Cryogenics*, 2006, 46, 794.
30. B.C. Ray, *Journal of Applied Polymer Science*, 2006, 102, 1943.
31. A. Nair and S. Roy, *Composites Science and Technology*, 2007, 67, 2952.
32. J. Ju, B.D. Pickle, R.J. Morgan, and J.N. Reddy, *Journal of Composite Materials*, 2008, 42, 569.
33. K.H. Lee and D.G. Lee, *Composite Structures*, 2008, 86, 37.
34. H.C. Hsia, C.C.M. Ma, M.S. Li, Y.S. Li, and D.S. Chen, *Journal of Applied Polymer Science*, 1994, 52, 1137.
35. T. Ueki, S. Mishijima, and Y. Izumi, *Cryogenics*, 2005, 45, 141.
36. T. Ueki, K. Nojima, K. Asano, S. Nishijima, and T. Akado, *Advances in Cryogenic Engineering*, 1998, 44, 277.
37. G. Yang, S.Y. Fu, and J.P. Yang, *Polymer*, 2007, 48, 302.
38. G. Yang, B. Zheng, J.P. Yang, G.S. Xu, and S.Y. Fu, *Journal of Polymer Science, Part A: Polymer Chemistry*, 2008, 46, 612.
39. A. Idesaki, N. Koizumi, M. Sugimoto, N. Morishita, T. Ohsima, and K. Okuno, *Advances in Cryogenic Engineering*, 2008, 54, 169.
40. R. Prokopec, K. Humer, and H.W. Weber, *Fusion Engineering and Design*, 2007, 82, 1396.
41. R. Prokopec, K. Humer, R.K. Maix, H. Fillunger, and H.W. Weber, *Fusion Engineering and Design*, 2007, 82, 1508.
42. C.J. Huang, S.Y. Fu, Y.H. Zhang, B. Lauke, L.F. Li, and L. Ye, *Cryogenics*, 2005, 45, 450.
43. J.P. Yang, G. Yang, G. Xu, and S.Y. Fu, *Composites Science and Technology*, 2007, 67, 2934.
44. J.P. Yang, Z.K. Chen, G. Yang, S.Y. Fu, and L. Ye, *Polymer*, 2008, 49, 3168.
45. S.Y. Fu, Q.Y. Pan, C.J. Huang, G. Yan, X.H. Liu, L. Ye, et al., *Key Engineering Materials*, 2006, 312, 211.
46. Z.K. Chen, G. Yang, J.P. Yang, S.Y. Fu, L. Ye, and Y.G. Huang, *Polymer*, 2009, 50, 1316.

47. E.W. Wong, P.E. Sheehan, and C.M. Lieber, *Science*, 1997, 277, 1971.

48. M.M.J. Treacy, T.W. Ebessen, and J.M. Gibsson, *Nature*, 1996, 381, 678.

49. M.F. Yu, O. Laurie, M.J. Dyer, T.F. Kelly, and R.S. Ruoff, *Science*, 2000, 287, 637.

50. M.F. Yu, B.S. Files, S. Arepalli, and R.S. Ruoff, *Physical Review Letters*, 2000, 84, 5552.

51. E.T. Thostenson and T.W. Chou, *Carbon*, 2006, 44, 3022.

52. F.H. Gojny, M.H.G. Wichman, U. Kopke, B. Fiedler, and K. Schulte, *Composites Science and Technology*, 2004, 64, 2363.

53. S.H. Joo, C. Pak, E.A. Kim, Y.H. Lee, H. Chang, D. Senney, Y.S. Choi, J.B. Park, and T.K. Kim, *Journal of Power Sources*, 2008, 180, 63.

54. D. Tasis, A. Tagmetarchis, M. Bianco, and M. Prato, *Chemical Reviews*, 2006, 106, 1105.

55. Z.K. Chen, J.P. Yang, Q.Q. Ni, S.Y. Fu, and Y.G. Huang, *Polymer*, 2009, 50, 4753.

56. B.C. Kim, S.W. Park, and D.S. Lee, *Composite Structures*, 2008, 85, 69.

57. P. Gupta, U.K. Scroop, G.S. Jhe, and M. Varime, *Polymer Plastics Technology and Engineering*, 2003, 42, 297.

58. M.R. Loos, L.A.T. Caelho, S.H. Pezzin, and S.C. Amico, *Materials Research*, 2008, 11, 347.

59. M.R. Loos, S.H. Prezzin, S.C. Amico, D.C. Bergmann, and A.A.I. Caelho, *Journal of Materials Science*, 2008, 43, 6064.

60. Q. Yan, W. Jiang, L. An, and R. Kyli, *Polymers for Advanced Technologies*, 2004, 15, 401.

61. J.K.W. Sandler, S. Pagel, M. Cedek, F. Gosny, M. Vants, J. Hochman, W.J. Blau, K. Schulte, A.H. Windle, and M.S.P. Shafer, *Polymer*, 2004, 45, 2005.

62. B. Wong, G. Gun, V. He, and J. Liu, *Polymer Engineering and Science*, 2007, 47, 1610.

63. V.E. Yudin, V.M. Svetlichnyi, A.N.S. Shumakov, and G. Maron, *Micromolecular Rapid Communications*, 2005, 26, 885.

64. S.M. Yeun, C.-C. Ma, C.-C. Ten, H.H. Wu, H.S.U. Cheng Kuang, and C.L. Chane, *Journal of Polymer Science, Part B: Polymer Physics*, 2008, 46, 472.

65. D. Wu, L. Wu, Y. Sung, M. Zhang, *Journal of Polymer, Part B: Polymer Physics*, 2007, 45, 3137.

66. Y.S. Song, *Polymer Engineering and Science*, 2006, 46, 1350.

67. K.Q. Xiao, L.C. Zhang, and I. Zaraudi, *Composites Science and Technology*, 2007, 67, 177.

68. H.-C. Kuan, C.-C. Ma, W.-P. Chang, S.-M. Yuen, H.-H. Wu, and T.-M. Lee, *Composites Science and Technology*, 2005, 15, 1703.

69. S. Peeterbroveck, M. Alexandre, J.B. Nagy, N. Morean, A. Destree, F. Monteverde, A. Rulmont, R. Jerome, and D. Dubois, *Micromolecular Symposia*, 2005, 221, 115.

70. H.C. Kuan, Private communication.

71. M. Naroozi and G. Zabarsad, *Journal Vinyl and Additive Technology*, 2010, 16, 147.

72. B. Hausrerova, N. Zdrajilova, and P. Saha, *Polymer*, 2006, 5, 33.

73. F. Barrero-Bujans, T. Armyo, E. Sa Juan, I. Rodrigues-Ramos, M. Pelez Cabero, and M.A. Lopez-Manchedo, Private communication.

74. N.K. Guimard, N. Gomez, and C.E. Schmidt, 2007, 32, 876.

75. M. Lai, J. Li, J. Yang, L. Liu, X. Tong, and H. Cheng, *Polymer International*, 2004, 53, 1479.

8 Applications of Reinforced Plastics in the Automobile Industry

A wide range of plastics are now being used in interior, exterior, and under-bonnet applications in the manufacture of automobiles.

Applications involving a range of plastics are reviewed in Table 8.1. It is seen in the table that the use of reinforced plastics is now penetrating the automotive industry very deeply, in many cases replacing the uses of metals. It can also be seen that glass fiber is probably the most widely used reinforcing agent.

8.1 POLYPROPYLENE

Blinkhorn et al. [1] developed a new glass fiber–reinforced polypropylene substrate for use in car interiors as a semistructural and acoustic material. It is available in a number of weights per unit area and with various proportions of glass. The composite is claimed to have exceptional expansion properties, which allow components of varying thickness to be molded.

Pohl and Miklas [2] have pointed out that one of the driving forces behind its acceptance by car designers is its nanomaterial construction. For example, the instrument panel parts can include long glass fiber polypropylene for the structure and talc-filled polypropylene for ductwork. One recent advance in long glass fiber polypropylene is MTX concentrates supplied by Verton. The heat-stabilized master hatch is 75% long glass fiber by weight and is available in 13 and 15 mm long pellets.

It is specifically formulated for machine-side blending using neat polypropylene. Verton MTX is based on commingled fiber technology called Twintex from St. Gobain. Pohl and Miklas [2] compared the performance properties of a fully compounded heat-stabilized 30% long glass fiber polypropylene to those of a 75% long glass fiber polypropylene master batch diluted to a comparable 30% glass fiber fill. These tests helped to verify that the performance properties are not compromised by the motor batch approach.

Engineering plastics such as polypropylene can be reinforced at the macroscopic level with a variety of higher-modulus materials such as fibers, beads, and cement, forming heterogeneous composites [3, 4]. Thus, reinforcing polypropylene with glass fibers as a cowoven mat has many advantages. Alternatively, reinforcement can also take place at the molecular level. These workers report on the benefits of simultaneous molecular and macroscopic reinforcement. Low fractions of polymer liquid crystal interlayers serve to increase the damping energy as well as increase

TABLE 8.1

Automotive Application of Reinforced Plastics

Plastic	Reinforcing Agent	Application
Polypropylene	20% talc	Automotive applications
Polypropylene	20% glass fiber	Automotive applications
Polypropylene oxide	10% glass fiber	Instrument panels
Polyether ether ketone	20% glass fiber	Automotive applications
Polybutylene terephthalate	10%–30% glass fiber	Distribution caps
Polyethylene terephthalate	35% glass fiber	Exterior body parts
	15%–55% glass fiber	Casings, housing, wiper arms
Polyethylene terephthalate	45% mineral and glass fiber	Automotive ignitions, roof liners
Polyamide 6	30% glass fiber	Automotive parts
Polyamide 6,6	50% mineral	Underbody parts
Polyamide 6,6	Carbon fiber	Connecting rods
Polyamide 11	30% glass fiber	Fan blades
Polyimide	40% glass fiber	Gear boxes
Polyetherimide	10%–30% glass fiber	Cooling fans, gears, underbody components, fuel systems
Thermoplastic polyolefin	Exfoliated clay	Timing belt (Toyota)
Polyphenylene	40% glass fiber	Head lamps, petrol, oil, end hydraulic fluid-resistant parts, exhaust emission control valves
Polyethylene sulfide	20% polytetrafluoroethylene lubricated	Antifriction gears
Polysulfone	10% glass fiber	Under-bonnet applications, electronic ignition components
Polysulfone	30% carbon fiber	Under-bonnet applications

the glassy plateau modulus. Long-term dimensional stability is mainly influenced by the presence of the woven glass fiber fraction in the cowoven mat. However, higher loadings of polymer liquid crystal interlayer have detrimental effects on the dimensional stability, causing interply slip at lower temperatures. Benefits of using a polymer liquid crystal interlayer ply between the cowoven mat are mainly associated with the weight savings to be gained by using a material with lower specific gravity. The crystallization kinetics of the composites are affected by the nucleating effects of the polymer liquid crystal interlayer and the glass fiber. When both components are present, the glass fibers dominate the nucleation kinetics. This is also reflected in the degradation kinetics, where complete degradation of the matrix occurs at lower temperatures for the glass-reinforced materials.

Magnetite has been used as a reinforcing agent for polypropylene [5]. Magnetite has good overall properties, such as thermal and electrical conductivity, ferromagnetism, and high density and hardness. Apart from these specific properties, it is particularly important in many applications, for example, in the automotive industry, for a filled plastic to have very good mechanical characteristics and to comply with recycling requirements. The trials showed that it is relatively easy to compound magnetite into a polypropylene matrix and injection mold the resultant compound.

Devareddy and Arbetti [6] have discussed the use of natural fiber composites with polypropylene in the automotive industry. They point out that due to their intrinsically superior properties, fibers may, in due course, replace glass fibers in may automotive applications.

8.2 POLYETHER ETHER KETONE

Winkler [7] and LNP Engineering [8] have also discussed the use of carbon fiber–reinforced polyether ether ketone in automotive applications. LNP Engineering [8] has developed a series of formulations of carbon aliphatic ketones based on thermoplastic compounds containing glass and polytetrafluoroethylene as a lubricant. These polyketones are targeted at the automotive market and are characterized by good impact performance over a broad temperature range. They have good chemical resistance and stability and superior resilience.

8.3 POLYBUTYLENE TEREPHTHALATE

Gears manufactured from fiber-reinforced polybutylene terephthalate have an extremely smooth surface and high maximum operating temperature up to 170°C. This composite is superior to polyacetal or metals and is often used in the manufacture of engine housings.

8.4 POLYETHYLENE TEREPHTHALATE

Glass fiber reinforcement of polyethylene terephthalate has been used extensively in automotive body moldings and bumper bar housings [9–11].

Brack et al. [9] found that these automotive components consisting of a metal insert overmolded with impact modified glass fiber–reinforced polyethylene terephthalate cracked adjacent to the insert following thermal shock testing (100 cycles –40°C to 180°C). The melt flow index of the material increased by 140% on drying prior to molding, suggesting a molecular weight reduction due to polymer degradation. However, the intrinsic viscosities before and after drying and after molding were similar, indicating minimal degradation. This anomaly was attributed to the melt viscosity, and intrinsic viscosity techniques are more sensitive to low-molecular-weight fractions.

8.5 UNSATURATED POLYESTERS

Krigbaum [12] has reported on the use of glass fiber–reinforced composites based on Cyglas 685 unsaturated polyester BMC and Cyglas 695 vinyl ester resin BMC (Cytec Industries) in automotive valve covers and other engine cover applications. He reviews development programs that led to the successful introduction of these components. The recyclability of thermoset composite valve covers is also discussed.

8.6 POLYAMIDES

Various polyamides, particularly polyamide 6,6 and polyamide 11, have been used in the manufacture of radiator tanks [13–15], rocker covers [16–22], Audi pedal boxes [23], connecting rods, fan blades, and other components [24]. Glass fiber reinforcement is used in these applications [17–24].

The requirements for radiator tanks are good heat stability, vibration resistance, and resistance to coolant additives [18].

Audi and others [20–22] use 25% glass fiber and 15% mineral reinforced nylon 6,6 for the fabrication of rocker covers for the Audi A4 and VW Passat engines.

These covers have good dimensional stability even at elevated temperatures, which ensures a low creep tendency that contributes to a long in-service life of the part.

8.7 POLYPHENYLENE SULFIDES

Forty percent glass fiber–reinforced polyphenylene sulfide is used in Ford automotive fuel rails [23]. This engine component conveys fuel from the injection pump to the cylinders and is required to withstand rapid changes of pressure, service temperatures of up to 120°C, engine vibrations, and direct contact with aggressive fuels. Details are given of chemical resistance tests carried out on the Fortran grade and also on a branched polyphenylene sulfide grade and two polyamide 6,6 grades. Test fuels are composed of ASTM reference fuel C, two aggressive fuels with differing methanol proportions, and an auto-oxidized fuel containing peroxide and copper additives. Fortran 1140L4 (a linear partially crystalline polyphenylene sulfide) exhibited the highest impact resistance and dimensional stability.

Polyphenylene sulfide offers high stiffness dimensional stability, and fatigue and good chemical resistance at temperatures as high as 200°C. It is finding broad use in demanding automotive applications.

8.8 NATURAL FIBER COMPOSITES FOR AUTOMOTIVE APPLICATIONS

Devareddy and Arbetti [6] have discussed this topic in detail. Stringent vehicular emission regulations and end-of-life vehicle (ELV) norms across the globe have mandated auto manufacturers to innovate highly recyclable and biodegradable plastic solutions. With biomimetics becoming more acknowledged in everyday products, researchers are trying to expand their horizon of overimplementation of biodegradable and recyclable materials in the automotive industry. Long-established materials such as wood and various natural fiber systems have been accepted as biodegradable materials. In-depth studies and experimental results have proven the replacement of traditional plastics and reinforced composite systems by natural fibers for engineering applications primarily in automotive interiors. Natural fibers may not perform in parallel with glass-reinforced systems in terms of mechanical properties, but they impart an environment-friendly nature to the overall system. These workers have discussed the application of natural fiber systems in automotive interiors.

The history and track record of the transportation industry shows an enlightening picture of the technological, scientific, and economic factors that changed environmental aspects in the world. There have been continuous efforts to develop newer alternative materials that achieve cost-effectiveness, improved fuel efficiency, reduced emissions, and increased safety while at the same time targeting recyclability and biodegradability. These efforts have been promoted, in part, by the increase in crude oil prices. While automakers have been talking up a future filled with biodiesel, hydrogen fuel cells, and even ethanol, they have been very quietly integrating parts made from composite materials into their present-day cars and trucks, thereby achieving reduced fuel consumptions and emissions. According to statistics, around a 25% reduction in the weight of a vehicle contributes to a savings of 250 million barrels of crude oil and a CO_2 emission reduction in the range of 100 billion tonnes per annum. A car manufactured today is much more advanced in terms of fuel consumption and safety features than a car made 25 years ago. The most proven practice to achieve this is to reduce the vehicle weight by incorporating composite-based structures. Glass-filled systems provide a wide range of applications in the automotive sector for structural applications.

Increased social awareness and environmental concerns posed by the nondegradable and nonrecyclable contents of salvaged vehicles are forcing automotive manufacturers to enhance the biodegradable content, which greatly favors switching over to natural fiber systems. To accelerate the process of switching to recyclable and biodegradable constituents, legislation in North America and Europe has issued specific directives over end-of-life vehicles that promote maximum use of environmentally safe products and reduce the landfill space requirement. This legislation has now come into effect, and it determines a deposition fraction of 15% for the year 2005 and 5% for the year 2015 [4].

In spite of high competition from advanced steel glass-reinforced plastic systems, natural fibers are finding significant use in vehicles and experiencing a steady growth. Other than the environmental and biodegradable advantages over glass filler composite systems, natural fibers are becoming more appreciated in the industry due to the following reasons:

- Comparable weight reduction of 10%–30% in plastic parts
- Good mechanical and manufacturing properties
- Good impact performance, high heat stability, and minimal splintering
- Occupational health advantages in assembly and handling compared to glass fibers, where airborne particles are proven carcinogens and can damage respiratory systems
- Recyclability of waste

REFERENCES

1. A. Blinkhoen, L. Barsotti, T. Cheney, E. Hague, and G. Knoll, *International Polymer Science and Technology*, 2006, 33, 1.
2. D. Pohl and M. Miklas, *Machine Design*, 2002, 20, 58.
3. S. Owen, D. Seader, N.A. D'Souza, and W. Broston, *Polymer Composites*, 1998, 19, 107.

4. W. Bristow, T.S. Dziemianowicz, J. Romanski, and W. Werher, *Polymer Engineering Science*, 1988, 28, 785.

5. P. Duifhuis and B. Weidenfeller, *Kunststoffe Plast Europe*, 2002, 92, 20.

6. S. Devareddy and M. Arbetti, *Popular Plastics and Packaging*, October 2002, p. 44.

7. E.P.J. Winkler, *Materials for Transportation Technology: Environment*, vol. 1, Wiley, Weinheim, 2000, p. 372.

8. LNP Engineering Plastics, *Plastics and Rubbers Weekly*, 1997, 1698, 13.

9. H.P. Brack, D. Ruegg, H. Buhrer, M. Slaski, S. Alkan, and G.G. Scherer, *Journal of Polymer Science, Part B: Polymer Physics*, 2004, 42, 2612.

10. R. Dweiri and J. Sahari, *Journal of Power Sources*, 2007, 171, 424.

11. S.A. Gursch, J. Schneider, H.B. Youcef, A. Wokaun, and G.G. Scherer, *Journal of Applied Polymer Science*, 2008, 108, 3577.

12. R.S. Krigbaum, *Composite Plastiques Reinforced Fibres de Verre Textile*, July/August 1997, pp. 22, 60.

13. R. Schouwenaers, S. Cerrud, and A. Ortiz, *Composites, Part A: Applied Science and Manufacturing*, 2002, 33A, 551.

14. Renault SA, *Plasticised and Rubber Weekly*, 1984, 1063, 25.

15. A.S. Baunin, T.V. Ponomareva, and O.B. Hunina, *International Polymer Science and Technology*, 2002, 29, T/52.

16. F.J.W. Wocof, J. Wolk, and C. Rhodia, *Engineering Plastics European Plastics News*, 2001, 28, 64.

17. Nyltech, *Plastics and Rubbers Weekly*, February 1997.

18. Rhudia Engineering Plastics, Woco, and F.J. Wolf & Co., *Plastics Additives and Compounding*, 2001, 3, 14.

19. M. Youson, *Engineering*, 1999, 240, 50.

20. Bayer AG. *Plastiques Flash*, 1998, 307, 70.

21. J. Linazisoro, *Plast 21*, 1996, 49, 50.

22. Radici Plastics, *Revista de Plasticos Modernos*, 2002, 83, 283.

23. S. Malek and B. Koch, *Renstade Plasticos Modernos*, 1999, 78, 164.

24. Ticona GMBH, Heochst A.G. Frankfurt am Main, 1999, p. 4.

9 Applications of Reinforced Plastics in the General Engineering Industry

9.1 MISCELLANEOUS ENGINEERING APPLICATIONS

In addition to specialized applications of reinforced plastics, as discussed in Chapters 7 and 8, plastics are used in a wide range of engineering applications, such as gears and bearings. Some of these applications are listed below.

General engineering applications are reviewed in Table 9.1. These include gears, bearings, valves, seals, pump impellers, piston rings, pipes and hoses, compression valves, plugs, and wear pads.

9.2 PLASTIC GEARS

One of the critical applications of reinforced plastics is in the manufacture of gears. Traditionally having been fabricated in metals, plastic gears are discussed in more detail below.

A wide range of reinforced plastics have been used in the manufacture of gears (see Table 9.1). The reinforcing agent used is usually glass fiber in the concentration range 10%–55% or carbon fiber at 30%.

The strongest growth area in the manufacture of gears has been in the automotive and aerospace applications. Automakers have sought to power a variety of vehicles systems with motors and gears. This has brought plastic gears into uses ranging from lift gates, seating, and tracking headlights to brake actuators, electronic throttle bodies, and turbo controls.

Appliances also make broad use of plastic power gears. Some larger applications, like washing machine transmissions, have pushed the limit on gear size, often as a replacement for metal. Plastic gears are present in many other areas, for example, damper drives in HVAC zone controls, valve actuators in fluid devices, automatic flushers in public restrooms, power screws that shape control surfaces on small aircraft, and gyro and steering controls in military applications.

Essential properties of gears are good fatigue properties, good wear resistance and lubricity, rigidity for high tear forces, and toughness for shock-loaded situations, such as reciprocating motors. These crystalline polymers must be molded hot enough

TABLE 9.1

General Engineering Applications of Reinforced Plastics

Used in Fabrication Of	Reinforcing Agent
A—Gear	
Polyoxymethylene	30% glass fiber
Polyoxymethylene	30% carbon fiber
Polyethylene terephthalate	30%–55% glass fiber
Polyamide 6,10	10%–30% glass fiber
Polyamide 11	30% glass fiber
Polyamide 12	30% glass fiber
Polyetherimide	10% glass fiber
Polyamide-imide	10% glass fiber
Polyethylene-tetrafluoroethylene	10% and 30% glass fiber
B—Bearings	
Polyoxymethylene	30% carbon fiber
Polyoxymethylene	30% glass fiber
Polyamide 6,10	10%–30% carbon fiber
Polyamide-imide	Graphite
Polytetrafluoroethylene	15% graphite
Polyethylene-tetrafluoroethylene	60% bronze
Polyethylene-tetrafluoroethylene	15% graphite
Polyethylene-tetrafluoroethylene	10%–30% glass fiber
Polyethylene-tetrafluoroethylene	30% carbon fiber
C—Valve Seals	
Polyamide 6,10	10%–30% glass fiber
Polyamide-imide	Graphite
Polyphenylene sulfide	30% carbon fiber
Polyphenylene sulfide	Glass fiber and glass beads
Polyphenylene sulfide	40% glass fiber
Polysulfone	30% glass fiber
Polysulfone	30% carbon fiber
Polyethylene fluoroethylene	Glass fiber
Polyethylene-tetrafluoroethylene	10%–30% glass fiber
Polyvinylidene fluoride	20% carbon fiber
Polyflourinated ethylene-propylene	20% glass fiber
D—Pump Impellers	
Polyether ether ketone	30% carbon fiber
Diallyl isophthalate	Long glass fiber
Polyamide-imide	Graphite
Polysulfone	30% glass fiber
Polysulfone	30% carbon fiber
Polyethylene-tetrafluoroethylene	10% carbon fiber

TABLE 9.1 (*Continued*)
General Engineering Applications of Reinforced Plastics

Used in Fabrication Of	Reinforcing Agent
E—Piston Rings	
Polyamide-imide	Glass fiber
Polyamide-imide	Graphite
Polyether fluoroethylene	60% bronze
Polyether fluoroethylene	15% and 25% glass fiber
Polyether fluoroethylene	15% graphite
Polyethylene-tetrafluoroethylene	30% carbon fiber
F—Pipes	
Polysulfone	30% carbon fiber
G—Valves	
Polyamide-imide	Glass fiber
H—Pumps	
Polyphenylene sulfide	Glass fiber and beads
Polyphenylene sulfide	40% glass fiber
Polysulfone	30% carbon fiber
I—Wear Pads	
Polytetrafluoroethylene	60% bronze
Polytetrafluoroethylene	15%–25% glass fiber
Polytetrafluoroethylene	15% graphite

to promote full crystallinity. Otherwise, gear dimensions can shift if the end-use temperature rises above the mold temperature and causes additional crystallization.

Some of the polymers meeting these requirements are polyacetals, polyethylene terephthalate, isobutylene terephthalate, polyamides 6,10, and 12, and polyphenylene sulfide.

Acetal has been a primary gear material in autos, appliances, office equipment, and other applications for over 40 years. It provides dimensional stability and high fatigue and chemical resistance at temperatures up to 90°C. It has excellent lubricity against metals and plastics.

Polybutylene terephthalate polyester produces extremely smooth surfaces and has a maximum operating temperature of 150°C for unfilled and 170°C for glass-reinforced grades. It works well against acetal and other plastics, as well as against metal, and is often used in housings.

Nylons offer great toughness and wear well against other plastics and metals, often in worn gears and housings. Nylon gears operate to temperatures of 175°C for glass-reinforced grades and 150°C for unfilled ones. Buy nylons are unsuitable for precision gears because their dimensions change as they absorb moisture and lubricants.

Polyphenylene sulfide offers high stiffness, dimensional stability, and fatigue and chemical resistance at temperatures as high as 200°C. It is finding broad use in demanding industrial, automotive, and other end uses.

Liquid crystal polymers offer great dimensional stability in small precision gears. They tolerate temperatures to 220°C and have high chemical resistance and low mold shrinkage. They have been molded to a tooth thickness of about 0.066 mm.

A thermoplastic elastomer helps gears run quieter and makes them more flexible and better able to absorb shock loads. A copolymer thermoplastic elastomer, for instance, is used in lower-power, higher-speed gears because it allows them to tolerate inaccuracies and reduce noise while providing sufficient dimensional stability and stiffness.

Material specification for gears and housings should take into account the dramatic effects fibers and fillers have on resin performance. For instance, when acetal copolymer is loaded with 25% short glass fiber (2 mm or less), its tensile strength more than doubles at elevated temperatures and its stiffness more than triples. The use of long glass fiber (10 mm or more) boosts strength, creep resistance, dimensional stability, toughness, rigidity wear, and other properties even further. This makes long fiber reinforcement attractive for use in large gears and housings to gain needed stiffness and better control of thermal expansion.

Polyethylene, polypropylene, and ultra-high-molecular-weight polyethylene have been used in gears at lower temperatures in aggressive chemical and high-wear environments. Other polymers have been considered for gears, but many impose severe limitations on gear function. Polycarbonate, for instance, has poor lubricity and resistance to chemicals and fatigue.

Acrylonitrile-butadiene-styrene terpolymer and low-density polyethylene generally cannot meet the fatigue endurance, dimensional stability, and heat and creep resistance requirements of precision gears. Such polymers are most often found in basic low-load or low-speed gears.

Liquid crystalline polymers such as thermoplastic elastomer copolymers are used in the manufacture of lower-power, high-speed gears because they allow the gears to tolerate inaccuracies and reduce noise while providing sufficient dimensional stability. Polyethylene, polypropylene, and ultra-high-molecular-weight polyethylene have been used in gears at lower temperature in aggressive chemical and high-wear environments.

Silicones and polytetrafluoroethylene have both been used to modify the properties of reinforced polymer gears.

Gamma sterilization does not affect the mechanical properties of polyether ether ketone and is well suited for human implantable devices [1]. Green et al. [2] explain why polyether ether ketone, an exceptionally strong engineering thermoplastic, is extremely well suited to use in long-term medical implant devices. Items discussed include polyether ether ketone qualities, manufacturing processes and biocompatibility testing, formulation with additives, carbon fiber compounds, carbon fiber composites, fiber reinforcement for load-bearing applications, and carbon fiber reinforcement for wear applications and medical applications.

TABLE 9.2

Applications of Lubricated Reinforced Plastics

Application	Polymer	Lubricant	%
Gears	Polybutylene terephthalate	Silicone	2
Gears	Polyamide 6,12	Silicone	2
Gears	Polyoxymethylene	Silicone	2
Gears	Polyoxymethylene	Polytetrafluoroethylene	—
Gears	Polyphenylene sulfide	Polytetrafluoroethylene	26
Gears	Polyamide 6,12	Polytetrafluoroethylene	20
Bearings	Polyoxymethylene	Polytetrafluoroethylene	—
Bearings	Polyimide	Polytetrafluoroethylene	—
Bearings	Polyoxymethylene	Silicone	2
Bearings	Polybutylene terephthalate	Silicone	2
Bearings	Polyamide 6,6	Molybdenum disulfide	—
Bushes	Polyoxymethylene	Silicone	2
Bushes	Polyamide 6,12	Silicone	2
Cogs	Polyamide 6	Molybdenum disulfide	—
Cams	Polyamide 6,12	Silicone	2
Valve seals	Polyamide 6,12	Silicone	2
Valve seals	Polyimide	Polytetrafluoroethylene	—
Valve seals	Polysulfone	Polytetrafluoroethylene	15
Valve seals	Polyimide	Graphite	25
Valve seals	Polyamide-imide	Graphite	—
Valve seals	Polyamide 6,12	Molybdenum disulfide	—
Pump impellers	Polysulfone	Polytetrafluoroethylene	15
Compression rings	Polyimides	Graphite	15

9.3   USE OF LUBRICATING AGENTS IN REINFORCED PLASTICS

Reinforced polymer grades are available that contain a lubricating agent such as silicone, graphite, polytetrafluoroethylene, or molybdenum disulfide. These impart special properties to the reinforced polymer, such as good wear characteristics and elongation at break performance. In a particular case, for example, the incorporation of 2% silicone into polyacetal improved its elongation at break from 6.5% to 56%. These properties are important in applications such as gears, valve seals, compression rings, bearings, and piston rings. Some examples of the use of lubricating agents are shown in Table 9.2.

REFERENCES

1. S. Green and K. Cartwright, *Proceedings of the Medical Polymers Conference*, Dublin, April 2–3, 2003, paper 5, p. 35.
2. S. Green, *Medical Device and Diagnostic Industry*, 2005, 27, 104.

10 Applications of Reinforced Plastics in the Aerospace Industry

Some applications of reinforced plastics in the aircraft and aerospace industries are listed in Table 10.1. It is seen in the table that many formulations have been used in aerospace applications that incorporate glass fibers or carbon fibers. For example, polyether ether formulations have been employed that use from 10% to 20% glass fibers, up to 40% in missile nose cones.

10.1 GLASS FIBER–REINFORCED PLASTICS: AEROSPACE APPLICATIONS

The aerospace industry has recognized the major benefits associated with fiber-reinforced composite materials [1]. The more popular techniques available for composite production are the traditional wet layup or autoclave and resin transfer molding. Efforts to further reduce processing time and improve part quality have focused on improved process control. To date, this has been based on off-line techniques. The need for on-line cure monitoring is widely recognized, and this will require the development of suitable in-mold sensors. The Engineering Composites Research Centre has investigated and concentrated on the specific problems encountered in the aerospace industry.

Buehler et al. [2] prepared prepreg materials from different suppliers, meeting the same aircraft materials specifications. While the specifications dictate the mechanical properties of the structures and some physical properties of the prepreg material, they place few requirements on handling and manufacturing behavior. Some parts, such as honeycomb structures, experience core crush, and several have a high level of porosity. These defects cause rejections of the parts and increased manufacturing costs. In addition, these parts can also result in some service-related problems, especially water ingression problems related to skin porosity. Accordingly, the main focus was to develop a better understanding of the physical characteristics of prepreg systems and to identify the appropriate evaluation procedures in order to reduce the number of reject parts.

Test methods used to evaluate these materials include resin content, gelation time, volatile content and resin flow resistance, prepreg tack, and friction. Additional evaluation techniques include Fourier transform infrared spectroscopy and thermal analysis.

Analysis of thermal transport phenomena in aerospace glass fiber–reinforced components, composite joints, and the composite material interface is very important. It has been shown by finite element analysis and molecular dynamics solutions that interface impedance plays an important role in dictating thermal transport

TABLE 10.1

Applications of Reinforced Plastics in the Aircraft Industry

Polymer	Reinforcing Agent	Percent	Applications
Epoxy	Kevlar prepreg	—	Aerospace applications, fuselage
Epoxy	Carbon fiber	—	Aerospace applications, helicopter blades
Polyether ether ketone	Glass fiber	20	Aerospace applications
Polyether ether ketone	Glass fiber	40	Aircraft and missile nose cones
Polyimide	Glass or carbon fiber	—	Parts of aero engines
Polyamide-imide	Glass or carbon fiber	—	Jet engine parts
Polyethersulfone	Glass fiber	30	Aerospace applications
Polysulfone	Carbon fiber	10	Aerospace applications
Polysulfone	Carbon fiber	30	Aircraft exterior and interior components
Polyvinylidene fluoride	Glass fiber	—	Film covering aircraft interior components

through the interface and that, through material modeling parameters, it can be identified for tailoring interface performance [5].

10.2   CARBON FIBER REINFORCEMENT PLASTICS AEROSPACE APPLICATIONS

Earlier work on the applications carried out up to 2002 was concerned with mechanical and thermal properties and ablation resistance. Thus, vapor-grown carbon fiber–reinforced phenolic resin composites have been evaluated for use in solid rocket motor nozzles.

Polymers with varying vapor-grown carbon fiber loadings in the range 30%–50% were exposed to a plasma lance for 20 s with a heat flux of 16.5 MW m^{-1} at about 1650°C. Low erosion rates and little char formation were observed, confirming that these materials had a promising use as rocket motor nozzle materials. The vapor-grown composites had low thermal conductivities (about 0.56 W m^{-1}), indicating that they were good insulating materials. If 65% fiber loading in vapor-grown carbon fibers could be achieved, then ablative properties would be expected to be comparable with, or better than, the composite currently used on the space shuttle reusable solid rocket motor.

Some 50 years ago, aircraft designers began taking advantage of the high strength-to-weight ratio associated with composites by replacing aluminum parts with others made from newer materials. In today's F-22 fighters, carbon composites compose nearly one-third of the jet's structure.

Composites have moved into terrestrial applications rather slowly. The primary obstacles have been the high cost of the materials and the labor-intensive operations and expensive fabrication equipment needed to process them. For various elements of the 500 hp Dodge Viper's fender support, door structures, and windshield frame, the carmaker chose parts made from carbon-based composites or blends of composites. Scientists at the Air Force Research Laboratory at Wright Patterson Air

Force Base are studying nanocomposites that exhibit unconventional combinations of properties [4–6].

McConnel [8] has pointed out that in the universe of complex high-performance aerospace components, carbon fiber epoxy composites have the ability to meet multiple functional requirements at prolonged service temperatures up to 121°C and can withstand short-duration spikes of 204°C. However, design goals for space vehicles, new commercial aircraft, and fifth-generation fighters have pushed sustained service temperatures into the range of 316°C or more, well beyond the capabilities of epoxies.

Five to ten years ago, these materials were still in the incubation period of development. Since that time, polymer research and system commercialization have progressed significantly. Desired weight benefits provided the primary impetus for producing cost-effective raw materials and developing qualified processes to ensure eventual production development. That development is now under way on flight-critical composite components in high-rate production.

Thirty years ago, the first polyimides were formulated in U.S. Air Force and NASA–funded programs. These polymerized monomeric reactant resins, based on aromatic dianhydride ester acid and aromatic diamine chemistry and polymerized through solvent addition, eventually became aerospace industry standard PMR-15 and PMR-11-50 formulations. Carbon fiber polymerized monomeric reactant resin composites outperform metals with greater strength-to-weight properties in aircraft engine nozzles and nacelles, helicopter gear cases, and missile fins when operating in the range of 450°F to 650°F (242°C to 342°C).

Although they are still in use today, polymerized monomeric reactant resins contain methylene dianiline, a hazardous compound and suspected liver carcinogen. Moreover, about 80% of the composite parts made with monomeric reactive resins are autoclave-cured with the balance compression molded. Several formulations have been developed as alternatives to monomeric reactive resins by scientists at NASA's Langley R&D Center (LaRC, Hampton, Virginia) and NASA Glenn (Cleveland, Ohio). One of these, RP-46, was patented in 1991, and offers similar chemistry but uses a different, less toxic diamine (3,4'-oxydianiline) to reduce toxicity.

RP-46 offers a glass transition temperature, T_g, of 740°F (393°C), and it demonstrates particularly high corrosion resistance in composites. An alternative to monomeric reactive resins, namely, DM B2-15, offered commercially by resin supplier and NASA-licensed Maverich Corporation (Blue Ash, Ohio), replaced monomeric reactive resin with 2,2-dimethyl-benzidine fiber composites as high as 635°F (335°C), and it can ensure service temperatures in carbon fiber composites as high as 335°C. NASA began development of another resin based on polyimide phenylethynyl-terminated polyimide (PETI) oligomers. These are being considered for hot-section military engine components presently under development.

Further improvements in polymerized monomeric reactants are achieved by end capping the phenylethynyl-terminated imide with resin precursor agents such as 4-phenylethynyl phthatic anhydride (PEPA) or biphenyl dianhydride–based polyimides, which offer toughness and flexibility up to 569°F (316°C) in carbon fiber nanocomposites.

FIGURE 10.1   Boeing Dreamliner.

Newer carbon nanocomposite polyimide formulations have even higher service temperatures, up to 1500°F (815°C), with flexural shear strengths up to 115 ksi at 600°F (316°C).

Loos [9] evaluated fiber uniweave fabric tackifier coating. This has been evaluated as a replacement material for tailored component aerospace structures [4, 9]. The fiber was impregnated with epoxy resin and oven-cured. Composite laminates that were fabricated using the modified process showed higher mechanical properties than the composite laminate fabricated using a traditional process.

In 2007, Boeing launched its high plastic Boeing 787 Dreamliner, in which preimpregnated carbon fiber was used for epoxy–matrix composite structural components [10] (Figure 10.1). Toughened polyacrylonitrile-based fiber and other polymers were also used in the construction. The 787 contains about 50% by weight of plastic components, including fuselage, wings, landing gear, horizontal stabilizers, passenger doors, wing tips, ice protection system, windows, nacelles, tail assembly, and hydraulic clamps.

Golfmans [11] has reported on an investigation of the cause of the crash of the American Airlines Airbus A300, Flight 587. The investigation showed that air turbulence was a possible factor in the crash.

This plane appears to have flown through the wake of a large plane that was much closer than originally thought. This led to a failure due to microbuckling, otherwise known as kinking, which generally occurs, in part, due to incomplete curing of the epoxy resin or to a deficiency in the hardening agent during the manufacturing process. Separately, the National Transportation Safety Board confirmed that the plane, which crashed after losing tail pieces and both engines, was the same Airbus A300 that ran into heavy turbulence in 1994. Structural experts checked whether the aluminum and plastic composite materials that were part of the tail assembly had been weakened by microscopic cracks and then broken by a later event. The purpose of this research was to investigate the dynamic stability of the leading/trailing panels and avoid parametric resonance. The leading/trailing panels, as part of the tail structures, take into consideration the fact that in the case of turbulence, left load will be significantly increased. In the case of loss of stability, the free and force vibration frequencies will occur at the same time.

10.3   PITCH FIBER CYANATE ESTER COMPOSITES

The *MESSENGER* (Mercury Surface, Space Environment, Geochemistry, and Ranging) spacecraft was the first to orbit the planet Mercury [12]. Designed and built by Johns Hopkins University's Applied Physics Laboratory (APL), the spacecraft orbited the planet for 1 year. In order to reduce the cost and schedule of this NASA Discovery mission, the solar arrays were required to be constructed of conventional, space-qualified materials. System thermal, mass, and stiffness requirements dictated that the panel facings be fabricated from a high-thermal-conductivity and -stiffness pitch fiber composite material capable of withstanding short-term temperatures as high as 270°C. A toughened, 177°C curing cyanate ester composite material resin system with extensive flight heritage was chosen, with a postcure used to extend the glass transition temperature closer to the maximum predicted temperature. A lengthy development program has been conducted at APL to provide assurance that the materials and processes chosen are capable of performing under such a demanding thermal environment. The results of this program will be applicable to other high-temperature spacecraft applications of advanced pitch fiber cyanate ester composite structures.

REFERENCES

1. A.T. McLehaggar, I.L. Haggar, S.T. Matthews, D. Brown, and B. Hill, *Proceedings of the ICAC 99 Conference*, Bristol, UK, September 23–24, 1999, p. 133.
2. F.U. Buehler, J.C. Seferis, and S. Zeng, *Journal of Advanced Materials*, 2001, 33, 41.
3. A.K. Roy, S. Siha, S. Ganjuli, and V. Varshariis, *Proceedings of the SAMPE Fall Technical Conference, Malfunctional Materials Working Together*, Memphis, TN, September 8–11, 2008, paper 44, p. 8.
4. Wright Patterson Air Force Base, *Advanced Materials and Processes*, 2002, 160, 13.
5. Proceedings of the 20th International SAMPE Technical Conference, Minneapolis, September 27–29, 1988.
6. R.D. Dalton, C.U. Pittman, L. Wong, J.R. Hill, and A. Day, *Composites, Part A: Applied Science and Manufacturing*, 2002, 33A, 243.
7. M. Jacoby, *Chemical and Engineering News*, 2004, 82, 34.
8. V.M. McConnel, *High Performance Composites*, 2009, 17, 39.
9. A.C. Loos, Dissertation, Mechanical Engineering Department, Michigan State University, East Lansing.
10. *Renstron Plastics News*, July 2, 2007.
11. Y. Golfmans, *Journal of Advanced Materials*, 2010, 42, 28.
12. P.D. Weinhold and D.F. Parsons, *Proceedings of the 34th International SAMPE Technical Conference, Materials and Processing Ideas of Reality*, Baltimore, MD, November 4–7, 2002, vol. 34, p. 308.

11 Radiation Resistance of Unreinforced and Reinforced Plastics

11.1 INTRODUCTION

Inasmuch as reinforcing agents such as glass fiber or carbon fiber are incorporated into polymers to improve their mechanical and other properties, any factor that might cause a deterioration in these properties is of interest in the applications of plastics in various industries. In some applications, particularly in the nuclear industry, medical instrumentation and packaging, microelectronics, and sterilization in the food industry, in which polymers are subject to gamma and other forms of radiation, the polymer can undergo damage often associated with a reduction in molecular weight. This, in turn, with some, but not all, polymers leads to serious deterioration in physical properties, such as impact and tensile strength. In an overall assessment of the risk of damage caused by gamma radiation, greater emphasis is now placed on changes in mechanical and other physical properties, and not, as was previously the case, the change in the color of the polymer.

Such considerations must be borne in mind when it is necessary to improve the physical properties of plastics, whether or not they have been reinforced by the addition of various agents. For example, a plastic that has been reinforced by the incorporation of glass fiber suffers a deterioration in physical properties. The greater the exposure, the greater is the loss in these properties.

As will be discussed below, the effects of gamma radiation are much more serious with some polymers than with others, several of which are highly resistant to the effects of gamma radiation. It is seen in Table 11.1 that polystyrene, polyether ether ketone, polyetherimide, and polyimides have an outstanding resistance to any damaging effect of gamma radiation. It is these polymers that are more likely to retain the improvement in their physical properties imparted by the incorporation of reinforcing agents in their formulations.

It is seen in Table 11.1 that commonly used polymers (e.g., polyolefins and their copolymers, polymethyl methacrylate) and polytetrafluoroethylene have very poor resistance to gamma irradiation and should not be used in applications in which exposure to this type of irradiation occurs.

While polyether ether ketone has an excellent resistance to the effects of gamma radiation, it is affected by exposure to protons and alpha particles [1].

Hafi et al. [2] studied the effect of 11 MeV protons and 26.6 MeV alpha particles on chain scission and cross-linking of polyether ether ketone.

Irradiation significantly reduced the thermal stability and durability of this polymer.

TABLE 11.1

Resistance of Polymers to Gamma Irradiation

Polymer	Degree of Resistance
PS	Excellent
PEEK	Excellent
PEI	Excellent
PI	Excellent
Cross-linked polyethylene	Very good
LDPE	Very good
Polyisobutylene	Very good
Styrene-butadiene	Very good
High-impact polystyrene	Very good
Epoxies	Very good
Polybutylene terephthalate	Very good
PET	Very good
Diallyl isophthalate	Very good
Diallyl phthalate	Very good
Alkyds	Very good
Ethylene tetrafluoroethylene	Very good
PC	Very good
Phenol formaldehyde	Very good
Styrene-maleic anhydride	Very good
Polyamide 11	Very good
Polyamide 6,6	Very good
Polyamide 6,10	Very good
Polyamide 6	Very good
Polyamide 6,9	Very good
Polyamide 12	Very good
Polyamide 6,12	Very good
Polyamide-imide	Very good
Polyurethane	Very good
Styrene-acrylonitrile	Very good
Acrylonitrile	Very good
Acrylate-styrene	Very good
PVDF	Very good
PVF	Very good
Chlorotrifluoroethylene	Very good
Chlorinated PVC	Very good
Unplasticized and plasticized PVC	Very good
Polyphenylene sulfide	Very good
Polysulfone	Very good
Polyethylene sulfide	Very good
Silicones	Very good
HDPE	Poor
PP	Poor
PMMA	Poor

TABLE 11.1 (*Continued*)
Resistance of Polymers to Gamma Irradiation

Polymer	Degree of Resistance
Ethylene-vinyl acetate	Poor
Urea-formaldehyde	Poor
PTFE	Poor
Petrafluoroalkoxy ethylene	Poor
Ethylene-propylene	Poor

PS = polystyrene, PEEK = polyether ether ketone, PEI = polyetherimide, PI = polyimide, LDPE = low-density polyethylene, PET = polyethylene terephthalate, PC = polycarbonate, PVC = polyvinyl chloride, PVDF = polyvinylidene fluoride, PVF = polyvinylidene fluoride, HDPE = high-density polyethylene, PP = polypropylene, PMMA = polymethyl methacrylate.

With regard to ultraviolet (UV) radiation, several polymers, reinforced or unreinforced, have excellent resistance to it. This includes polyether ether ketone, diallyl isophthalate, polyetherimide, vinylacetate, silicones, and several fluorinated polymers, such as polyvinyl fluoride, polyvinylidene fluoride, perfluoroalkoxy ethylene, ethylene-chlorotrifluoroethylene, and fluorinated ethylene-propylene copolymer (Table 11.2).

Several polymers combine excellent ultraviolet resistance with good tensile and elongation at break properties (Table 11.3). The storage modulus, alpha relaxation, and creep in polymers are influenced by electron irradiation. Thus, the creep of some polymers increased upon exposure to electron beam irradiation below 4 Mrad. Neutron/gamma irradiation also had an adverse effect on some polymer properties. Thus, some glass fiber–reinforced plastics lose 20%–40% of their flexural strength after exposure to neutron/gamma irradiation doses above 1×10^8 Gy [3].

A discussion now follows on the effect of various types of radiation on the retention of physical properties in reinforced and unreinforced polymers. We will start with a discussion of the effects of gamma radiation.

11.2 EFFECT OF GAMMA RADIATION ON THE PHYSICAL PROPERTIES

11.2.1 UNREINFORCED PLASTICS

We will start with a consideration of a plastic that has excellent resistance to gamma radiation relating to its physical properties, namely, polyimide.

11.2.1.1 Unreinforced Polyimides

Marinovic-Cincovic et al. [4] have studied the thermal, oxidative, and radiation stability of various imides based on *N*(3-(2.5 dioxo-2,5-dihydro-1*H*-pyrol-1-yl) phenyl)acetaimide of different diamines, namely, urea (designated AI), 6-amino

TABLE 11.2
Ultraviolet Radiation Resistance Polymers

Excellent	Very Good	Poor
Polyether ether ketone	Cross-linked	HDPE
Diallyl isophthalate	Polyisobutylene	LDPE
Ethylene-vinyl acetate	Styrene-ethylene-styrene	PP
Polyetherimide	Polystyrene	Polyisobutylene
Polyvinyl fluoride	Epoxies	Polymethylpentene
Perfluoroalkoxy ethylene	Diallyl phthalate	Ethylene propylene (E/P)
Ethylene chlorotrifluoroethylene	Ethylene-tetrafluoroethylene	Styrene-butadiene
Fluorinated ethylene-propylene	Polycarbonate	Polyoxymethylene
Silicones	Phenol formaldehyde polyamides 6; 6,6; 6,9; 6,10; 10; 12	Styrene-maleic anhydride
	Polytetrafluoroethylene	Polyamide-imide
	Chlorinated PVC	Polyimide
	Unplasticized and plasticized PVC	Polyurethane PMMA
	Polyphenylene sulfide	Urea formaldehyde
	Polysulfone	Styrene-acrylonitrile
		Acrylonitrile-butadiene
		Styrene
		Polyethersulfone

TABLE 11.3
Selection of Plastic with Excellent UV Resistance and Tensile and Elongation Properties

Polymer	Elongation of Break, %	Tensile Strength, MPa	UV Resistance
Polyether ether ketone	400	92	Excellent
Polyethylene terephthalate	300	55	Excellent
Ethylene-vinyl acetate	750	17	Excellent
Cross-linked polyethylene	350	18	Excellent
Perfluoroalkoxy ethylene	300	29	Excellent
Polycarbonate	200	50	Excellent
Ethylene fluoroethylene	200	30	Excellent
Fluorinated ethylene propylene	150	14	Excellent

hexylamine (designated AII), 2-amino ethylamine (designated AIII), 3/4-(3-amino-propyl)piperasinc-1-yl), propan-1-amine (designated AIV), 4/4(4-aminocyelohexyl) methyl cylcohexamine(designated AV), 4(4-aminobenzil)aniline (designated AVI), and 4(4-aminophenyl sulfonyl)aniline (designated AVII).

Thermal and thermooxidative behavior were studied by thermogravimetric analysis in oxygen and nitrogen. Polyimide resins were irradiated with a 500 kGy total adsorbed dose of gamma radiation.

Cobalt-60 gamma radiation and its radiation stability were evaluated on the basis of the thermal and thermooxidative behavior of the irradiated samples. Characteristic thermal degradation temperatures of unirradiated polyimides thermally treated in a nitrogen atmosphere were expressed as 5.10 and 30% mass loss 10.5%, T 10% and T 30%, respectively.

In the initial stage of decomposition, according to the initial decomposition temperature, the polyimide N-(3-(2·5-dioxo-2,5-hydro-1-1-1-pyrrole-1-yl)phenylacetamide bonded with 4(-4-aminocyclohexyl-)methyl-cyclohexan amine has the best thermal stability (initial decomposition temperature T of 250°C); the corresponding temperatures for 10% and 30% weight loss are 406°C and 457°C, respectively.

The generally accepted fact is that polyimides are radiation-stable materials. By analyzing the thermal stability of irradiated polyimide samples, we can conclude that their thermal stability, compared to unirradiated samples, is about the same or even better after irradiation.

Marinovic-Cincovic et al. [4] concluded that the thermal and oxidative degradation of the polyimides are complex and occur in two or more phases.

The thermal and thermooxidative stabilities depend on the cross-linking density, depending on the relative reactivity of the diamine used. The use of diamines, which have smaller or more flexible molecules, leads to the formation of more stable polyimides with higher cross-linking density.

The structure of the diamine component has a greater effect on the (nonoxidative) thermal than on the thermooxidative stability.

An applied dose of 500 kGy gamma radiation contributed to the better stability, due to additional cross-linking occurring in polyimides that were obtained from diamines, which contain the longest aliphatic chain length and aromatic, cyclo-aliphatic, or heterocyclic rings. The improvement of stability is the greatest for the polyimide **A-VI**, which was synthesized from N-[13-2.5-dioxo-2,5-dehydro-1-1-1(pyrroleyl)phenyl]acetamide 4-(4-aminobenzyl)aniline.

11.2.2 REINFORCED POLYIMIDES

Homrighausen et al. [5] evaluated the gamma radiation performance of glass fiber–reinforced polyimide laminates.

The polyimide resin formulations were based on combining various amounts of 3,4¹ oxydianiline and 4,4-(1,3-phenylenedioxy)dianaline with 4,4¹ bisphenol A dianhydride and 3,3¹,4,4¹-benzopehnone tetracarboxylic dianhydride and terminated in 4-phenylethynyl phthalate anhydride.

These glass fiber–reinforced 4-phenylethynyl phthalic anhydride–terminated poly-imides were of interest as possible S-2 glass fiber–reinforced composites as super-conducting magnetic material insulation intended for use in future nuclear fusion devices, such as the International Thermonuclear Experimental Reactor (ITER), Fusion Experimental Reactor (ITER), or Fusion Ignition Research Experiment (FIRE).

The effects of three different cobalt 60 gamma irradiation doses between 267 and 641 rad L^{-1} on cryogenic (77°K) mechanical and thermal properties were measured.

Woven-2-glass fiber–reinforced composites prepared from 3,4-oxydianiline doses of 25, 50, and 100 Mrad of gamma radiation at 77°K affected shear strength for two glass fiber–reinforced polyimides prepared from 3,4^1 dianiline (polyimide 3) and a bis-maleimide (RTM 65). The short-beam shear strength at 77°K of RTM 65 without fiberglass reinforcement falls from 87.1 to 79.3 MPa upon increasing the radiation dose from 0 to 1000 kGy. With polyimides, however, the short-beam shear strength at 77°K falls only from 92 to 88.9 MPa when the radiation dose is increased from 0 to 1000 kGy.

With unreinforced RTM 65, flexural strength at 77°K falls from 1227 to 1005 MPa when the radiation doses is increased from 0 to 1000 kGy, while with the fiberglass-reinforced polyimide the flexural strength falls by a much smaller amount, 1019 to 908 MPa, upon increasing the radiation dose from 0 to 1000 kGy.

With unreinforced RTM 65, the bis-maleimide flexural modulus falls from 27.1 to 0 GPa for zero exposure to gamma radiation after a radiation dose of 250 kGy. Under the same conditions, the flexural modulus of fiberglass-reinforced polyamide 3 increases from 19.9 to 25.3 GPa upon exposure to 250 kGy gamma radiation.

T_g measurements from polyimide 3 were essentially unchanged over a radiation dose of up to 1000 kGy, while the T_g values for bis-polyimide were reduced by 10% from the zero dose level. In all cases, the polyimide was found to be more resistant to radiation-induced damage than the bis-maleimide.

We will now move on to a polymer, namely, polyethylene (PE), that is reported to have a very good, as opposed to an excellent, resistance to the effects of gamma radiation.

11.2.3 UNREINFORCED HIGH-DENSITY POLYETHYLENE AND LOW-DENSITY POLYETHYLENE

Cota et al. [6] determined changes in the mechanical properties with the dose rate of gamma radiation. This study was concerned with high-density polyethylene. Consideration was given to the influence of the parameters on mechanical strength properties as a result of the predominance of oxidative degradation or cross-linking. The results showed an improvement in the mechanical strength of high-density poly-ethylene, with increased doses of gamma radiation indicating the predominance of cross-linking over oxidative degradation. Gamma radiation at low dose rates affects the mechanical properties of high-density polyethylene more efficiently than at high dose rates.

Analysis of the change in the mechanical properties of high-density polyethylene submitted to gamma irradiation shows that a 5% offset affects yield stress, and yield stress obtained from compression tests increased as the adsorbed radiation dose increased. This effect was more pronounced at lower dose rates. Most of the modulus of elasticity values for the compression tests were higher than those for the nonirradiated samples, although tendencies were not well identified for the lowest dose rate. Yield stress for the tensile tests decreased with an increase in dose, but for higher dose rates, this parameter began to increase. Also, the modulus of elasticity for the tensile tests decreased over all doses and dose rate intervals, and this effect was more pronounced at lower dose rates.

It is important to observe that the experimental framework used in this work was designed to allow a comparative study of the mechanical strength properties of nonirradiated and irradiated samples of a specific high-density polyethylene material. Thus, the validity of the data and conclusions obtained is limited by the assumptions and material used. In order to produce more general data, it is desirable to make use of other complementary tests to better characterize the changes in the microstructure of the material due to the application of gamma radiation.

Considering the experimental data from a comparative perspective, the results show an improvement in the high-density polyethylene mechanical strength properties as dose increases, indicating the predominance of cross-linking over oxidative degradation, despite the irradiation in air, which would theoretically ensure oxygen availability across the sample. These results indicate that sample thickness could be a very important parameter for the design of an irradiation test and should represent the real dimensions of the polymeric component to be submitted to radiation. The results also show that lower doses are necessary to obtain a similar change in mechanical strength parameters when radiation is applied at lower dose rates, showing that gamma radiation affects the high-density polyethylene in a more efficient way at lower dose rates.

Luyt [8] also reported that gamma radiation improved the thermal stability of low-density polyethylene. It also influenced Young's modulus and elongation at break, but doses of 150–200 kGy also caused crazing.

Kaci et al. [14] reported a loss of tensile properties of metallocene linear low-density polyethylene upon exposure to gamma radiation.

Kim et al. [15] showed that due to cross-linking, gamma irradiation of polyethylene improved the thermal shrinkage resistance of polyethylene battery separation materials compared to the thermal shrinkage of unexposed separations.

Babovich et al. [7] examined gamma-irradiated cross-linked polyethylene film for creep and stress relation characteristics. Increases in creep strains were observed for radiation doses exceeding Mrad. This was due to disorientation of the amorphous phase because of crystallinity.

11.2.4 REINFORCED HIGH-DENSITY POLYETHYLENE

11.2.4.1 High-Density Polyethylene (HDDE)–Alumina Composites

Dao-Long et al. [9] reported that the mechanical properties of these composites are improved significantly over these of unreinforced high-density polyethylene. Also, increasing the alumina content improves the thermal conductivity of the polymer.

11.2.4.2   HDDE–Hydroxyapatite Composites

Karam et al. [10] reported the effect of 25 kGy gamma radiation on the mechanical properties of high-density apatite composites. These workers compared mechanical and thermal properties of unirradiated and irradiated samples. They concluded that these composites are potential candidates for biomechanical applications.

11.2.4.3   High-Density Polyethylene–Ethylene Vinyl Acetate–Clay Nanocomposites

Lu et al. [11] showed that these nanocomposites had a superior resistance to gamma rays compared to those of unreinforced polymer. A plot of tensile strength versus gamma dose for polymers containing between 0% and 10% organically modified clay shows that below breakdown levels of 50 kGy, the remaining tensile strength was virtually unaffected by gamma radiation dose. Above this level of exposure to gamma radiation, the tensile strength increases appreciably, especially at higher clay contents. The variation of elongation at break of the nanocomposites with irradiation dose is less than that of polymer containing no clay. When exposed to gamma radiation, the elongation at break increases with initial doses up to 100 kGy and then decreases for all the samples examined.

11.2.5   UNREINFORCED ULTRA-HIGH-MOLECULAR-WEIGHT POLYETHYLENE

Suarez et al. [12] investigated the effect of gamma radiation on the mechanical properties of ultra-high-molecular-weight polyethylene. The nonirradiated materials have a high tensile strength and elongation at break, and the tensile properties decrease with exposure to gamma radiation.

11.2.6   REINFORCED ULTRA-HIGH-MOLECULAR-WEIGHT POLYETHYLENE

Xiang et al. [13] have studied the effect of gamma radiation on the properties of titanium dioxide–reinforced ultra-high-molecular-weight polyethylene.

11.2.7   UNREINFORCED POLYPROPYLENE

Unlike polyimides, as discussed earlier, polypropylene (PP) undergoes severe deterioration in mechanical and other properties when exposed to gamma radiation.

Hassan et al. [16] studied the effect of gamma irradiation on the thermal and mechanical properties of highly crystalline polypropylene.

Standard radiation doses of 20, 40, 60, 80, 100, and 120 kGy were applied to the samples of propylene.

The effect of gamma irradiation on the thermal characteristics of unreinforced highly crystalline polypropylene was studied by determination of melt temperature (T_m), melting enthalpy (ΔH_m), and degree of crystallinity values. All these values decrease with increasing radiation doses, indicating that the polymer becomes more amorphous during irradiation.

For example, an increase in radiation dose from 0 to 120 kGy decreased T_m from 126.6°C to 124.5°C, caused a decrease in enthalpy.

The differential scanning calorimetric curve of highly crystalline polypropylene at various gamma radiation doses between 0 and 120 kGy indicates an increase in thermal stability of the polymer with increasing radiation dose.

The unirradiated polypropylene sample remains stable from 30°C to 253°C, where there is no weight loss of the sample. This stable zone is followed by a slow rate of decomposition from 253°C to 400°C. A faster rate of decomposition starts between 400°C and 470°C, followed by a slower rate of decomposition between 470°C and 590°C, at which the sample is completely decomposed. A weight loss of about 94% has been recorded in this zone. The irradiated polymer remains stable up to a comparatively higher temperature, that is, from 30°C to 276°C for gamma irradiation at the 20 to 120 kGy level. It also follows the same trend of double-step decomposition as that of the unirradiated polypropylene. A slower rate of decomposition takes place from 270°C to 420°C where the sample loses about 10% of its initial weight. This is followed by a fast decomposition zone from 420°C to 460°C, with a weight loss of about 90%. The irradiated sample is completely decomposed at 570°C.

Stress–strain curves of irradiated highly crystalline polypropylene exposed to gamma rays exhibit its ductile behavior with elongation values for the nonirradiated samples up to 80%. After irradiation, highly crystalline polypropylene becomes less ductile and tensile stress decreases as well.

The E-modulus of crystalline polypropylene reflects the tensile behavior of the material under a minute deformation and virtually static test conditions, that is, in the ideal elastic region. The E-modulus of both gamma and electron beam–irradiated highly crystalline polypropylene rises with increasing irradiation doses up to 60 kGy, and then it decreases due to degradation of the main chain of highly crystalline polypropylene. Taking into account the phase composition of the specimens, this rise manifests an improvement of structure order, that is, with rising gamma-irradiated highly crystalline polypropylene. Considering the poorer thermodynamic stability of electron beam–irradiated highly crystalline polypropylene than that of gamma-irradiated highly crystalline polypropylene, one can assume that the E-modulus of the specimens irradiated with electron beams should be systematically lower than that of the specimens irradiated with gamma radiation.

Hassan et al. [16] also prepared plots of tensile strength at break against the irradiation dose for both gamma and electron beam radiation. It is observed that the tensile strength at break rises monotonically in the dose range of up to 60 kGy, and then it decreases. In the dose range 0–60 kGy, the highest rate of the increase is observed, equal to 0.02 MPa/kGy for samples irradiated by gamma rays and 0.04 MPa/kGy for those irradiated by electron beam on average irradiation dose increases.

It is concluded that the thermal properties (melting temperature, melting enthalpy, and degree of crystallinity), thermal stability, and mechanical properties (elastic modulus tensile strength and extension at break) of highly crystalline polypropylene showed change after irradiation.

Stojanovic et al. [17] studied the effect of high doses of gamma radiation up to 700 kGy on the melting behavior of uniaxial isotactic polypropylene. These workers showed that with both unoriented and oriented polypropylene, increasing radiation also reduced both heat of fusion and melting temperature (T_m).

Beyond a radiation dose of 250 kGy, gamma radiation decreases crystallinity as a consequence of gel formation. Melting temperature decreases with increasing absorbed gamma radiation dose.

Nedkov and Dabreva [18] showed that in isotactic polypropylene, radiation doses up to 100 kGy improve crystalline perfection, but with high gamma radiation doses up to 200 kGy, damage occurs in both the crystal and amorphous laminar parts of the polypropylene.

11.2.8 REINFORCED ISOTACTIC POLYPROPYLENE TITANIUM DIOXIDE AND OTHER COMPOSITES

Mina et al. [19] demonstrated improvements in performance in isotactic polypropylene containing titanium dioxide filler. These include an increase in microhardness and impact properties with an increase in titanium dioxide content in the composite. Also, thermal stability is enhanced and electrical resistivity decreases with an increase in titanium dioxide content.

In recent years, polymer-layered silicate nanocomposites have attracted considerable attention in both fundamental research and industry and have been considered a new generation of composite materials. They often exhibit excellent mechanical, thermal, and gas barrier properties, and it is predicted that these materials will play an important part in industries as diverse as automotive, packaging, foodstuffs, and aerospace.

Tonato et al. [20] studied the effect of up to 100 kGy gamma radiation on the properties of organically modified montmorillonite clay-propylene nanocomposites. These workers reported that oxidative degradation under gamma irradiation of polypropylene-clay nanocomposites causes drastic modification in the structure, morphology, and tensile and thermal properties of the nanocomposites, especially at doses above 20 kGy.

A large increase in carbonyl index was noted, accompanied by a significant decrease in melting temperature (T_m) index and ultimate tensile properties.

The oxidation rate of polypropylene nanocomposites upon gamma irradiation is much higher than that of polypropylene.

The Young's modulus of neat polypropylene remained fairly constant at about 7200 MPa, whereas for the polypropylene-clay nanocomposite, it fell from a value of 1400 MPa for zero exposure to gamma radiation to about 1200 MPa after exposure to 30 kGy of gamma radiation. Thus, the increase in Young's modulus of 1200 to 1400 MPa brought about by incorporating clay in the formulation is lost after exposure to 30 kGy gamma radiation, when Young's modulus falls to 1200 MPa. Residual stress at break values for both neat polypropylene and polypropylene-clay nanocomposite decrease dramatically after exposure to gamma radiation above 16 kGy.

11.2.8.1 Perlite-Filled Polypropylene

Akin-Okten et al. [21] studied the effect of up to 1000 kGy gamma radiation on the mechanical properties of perlite-filled polypropylene containing up to 50% perlite.

It was found that the ultimate tensile strength, elongation, and impact strength of the composites all decreased with increasing gamma irradiation. Yet these changes appeared not to occur faster than the change in unfilled polypropylene upon irradiation.

The stress at break decreases almost half that of the nonirradiated polypropylene after exposure to 50 kGy at doses of 10 and 25 kGy; the stress at break for 15% perlite-polypropylene composites had almost the same value as that of virgin unfilled polypropylene. The variation elongation with increase in dose was found to be even faster than the stress at break after 25 kGy gamma irradiation. The addition of perlite into polypropylene still appeared to give acceptable tensile strength values of 50–100 kGy for irradiated polypropylene for exposure to 50–100 kGy gamma radiation when the perlite content was 15% and 30% by weight, yet at these high doses, the loss in strength is inevitable.

The percentage elongation at break values showed a decreasing trend with the irradiation dose and with the filler content. Perlite addition to nonirradiated polypropylene caused a sharp drop in elongation at break, from nearly 8% elongation for unfilled polypropylene to 2.5% at 50% perlite addition. However, in irradiated polypropylene, the dose of irradiation for a given perlite content was not very large, especially at the 50% perlite level.

The addition of perlite into virgin unfilled polypropylene reduced the impact strength from 20 kN/m (0.6 J) to a value less than 5 KN/m (0.15 J). The impact strength of unirradiated unfilled polypropylene also showed a very fast decrease, but it remained almost unchanged with the absorbed gamma radiation dose, with the exception of the 25 kGy dose.

The effect of gamma radiation dose on melting temperature (T_m) for polypropylene containing between 0% and 50% perlite was studied. The melting temperature of the unfilled polypropylene decreases with dose of radiation. This decrease becomes even higher by the addition of perlite into gamma-irradiated polypropylene compared to the unfilled polypropylene with nonirradiated polypropylene, where the addition of perlite increased melting temperatures from 165°C to 172°C.

In conclusion, it was found by Akin-Okten et al. [21] that the ultimate strength of perlite-filled polypropylene after a 25 kGy gamma radiation dose was better than that of unfilled irradiated polypropylene.

In general, the tensile strength was found to be improved with perlite addition at higher radiation doses, especially for 15% and 30% perlite compositions. The ultimate elongation also decreased with addition of perlite and the radiation dose, but not as fast as observed in unfilled polypropylene, especially when 50% perlite composites are considered. The impact strength, as can be expected, decreased with perlite addition for nonirradiated polypropylene composites, but it was found that gamma irradiation for both unfilled polypropylene and filled polypropylene did not much alter the impact strength.

Thus, the addition of mineral filler perlite apparently improves the mechanical properties of gamma-irradiated polypropylene. Even at a high dose of irradiation, perlite addition resulted in substantially acceptable mechanical strengths, indicating better interfacial interaction than the variation of the mechanical properties in non-irradiated polypropylene composites and the unfilled polypropylene.

Studies have also been conducted on the effect on mechanical properties of polypropylene filled with talc [22].

11.2.9   EFFECT OF GAMMA RADIATION ON OTHER POLYMERS

As might be gathered, published work now exists on the effect of gamma radiation on a wide range of polymers, whether they be in their virgin state, that is, without reinforcing agents (Table 11.4) or fillers, or contain one of a variety of reinforcing agents or fillers (Table 11.5).

Enough has been said to show that (1) gamma radiation affects polymer properties, beneficially or otherwise, and (2) in some cases, the incorporation into the polymer formulation of a reinforcing agent or filler causes an improvement, and sometimes causes a deterioration of physical properties when the polymer is exposed to gamma radiation.

11.3   EFFECT OF ELECTRON IRRADIATION OF PLASTICS

11.3.1   POLYPROPYLENE

Hassan et al. [16] studied the affect of electron and gamma beam irradiation on the thermal and mechanical properties of highly crystalline polypropylene.

The effects of the structural modifications of highly crystalline polypropylene that could occur following these treatments at different doses (20, 40, 40, 80, 100, and 120 kGy) were studied. These were applied to all samples and compared to the properties of the unirradiated sample. Irradiation of samples of highly crystalline polypropylene decreased the melting temperature (T_m) of polypropylene; it also decreased the melting temperature after exposure. This decrease is up to 5°C. The changes in mechanical properties exhibit different radiation stability toward 60 Co gamma radiation and electron beam irradiation. This difference reflects a much higher penetration of the gamma radiation through the polymeric material as a function of sample thickness. The degradation of polymer properties caused by gamma irradiation was more than that caused by electron beam irradiation.

Krauz et al. [61] studied the effects of temperature during the electron beam irradiation of isotactic polypropylene. These workers noted a slight reduction in molar mess, an increase in long-chain branching, and an increase in crystallization temperature with increasing temperature.

Radwen [62] investigated the effect of electron beam irradiation in the optical properties of polypropylene.

11.3.2   LOW-DENSITY POLYETHYLENE

Babovich et al. [63] studied creep and stress relaxation of low-density polyethylene film cross-linked with beta radiation as a function of irradiation dose. They showed that both storage modulus and alpha relaxation were influenced by irradiation. Creep results showed an increase in the creep strain when the polymer was irradiated with a dose below 4 Mrad, in comparison with a nonirradiated low-density polyethylene. This increase corresponded to a disorientation in the amorphous phase, which took place as a result of the film heating during irradiation.

TABLE 11.4
Literature on the Effects of Gamma Radiation on Unreinforced Polymers

Polymer	Range of Gamma Radiation Dose, kGy Unless Otherwise Stated	Properties Studied	Reference
Ethylene octane copolymer, hyperbranched low-density polyethylene (VG)	0·2–10 mGy	Mechanical thermal	23
Polycarbonate (VG)	3·5–10 Mrad	Thermal	24
Polymethyl methacrylate (P)	—	Mechanical	25
Polymethyl methacrylate doped with benzylidene polymer	—	Structural	26
Polymethyl methacrylate (dithizone and carbazone couplexes)		Molecular weight, crystallinity	27
Polymethyl methacrylate	Up to 100	Environmental stress cracking, elongation	28
Polymethyl methacrylate	Up to 100	Mechanical properties, stress cracking	29
High impact polystyrene (VG)	—	Fracture toughness	30
Styrene/butadiene/rubber (VG)–toughened polystyrene (E)	Up to 250	Mechanical	31, 32
Polystyrene-polypropylene blend	Up to 70	Tensile properties	33, 34
Styrene-butadiene acrylonitrile	—	Mechanical	52
Polybutylene (VG)	—	Degradation	53
Polybutadiene	8–23	Molecular weight Thermal	54, 55
Polystyrene	Up to 100	Color	34
Polybutylene terephthalate/ co-adipate (VG)	Up to 300	Thermal and mechanical	35
Polyvinyl butyral	Up to 300	Mechanical thermal	36
Ethylene-propylene-diene High-density polyethylene-rubber blends	Up to 300	Mechanical thermal	37
Ethylene-propylene elastomers	—	Thermal	38
Poly(3-hydroxybutyrate)	Up to 180	Mechanical thermal	39
Polyether ether ketone (E)	—	Molecular weight, mechanical	40
Polyethylene terephthalate (VG) Polyethylene Polytetrafluoroethylene Polycarbonate Polyimide (E)	—	Electrical	41
Polyetherimide (E)	—	Electrical	42
Polyacrylamide	—	Mechanical	43

Continued

TABLE 11.4 (*Continued*)

Literature on the Effects of Gamma Radiation on Unreinforced Polymers

Polymer	Range of Gamma Radiation Dose, kGy Unless Otherwise Stated	Properties Studied	Reference
Polyarlyonitrile	Up to 120	Mechanical thermal, electrical	44
Polyurethanes (VG)	25	Mechanical	45
Polymurethanes	10^5 rad L^{-1}	Mechanical	46
Triallisocyanurate	Up to 100	Mechanical thermal	47
Polyamide 6,12 (VG)	—	Thermal	52
Polytetrafluoroethylene	—	Thermal	48
Fluoropolymers	—	Thermal chemical stability	49
Polyvinylidene difluoride (VG)	0–300 kGy	Optical chemical	50
Polyvinylidene fluoride-polymethyl methacrylate	25 Mrad	Radiation hardening	51
Polysulfones	Up to 100	Chemical effect	56
Silicones (VG)	Up to 50 Mrad	Mechanical	57

E = excellent resistance to effects of gamma radiation, VG = very good resistance to gamma radiation, P = poor resistance to gamma radiation.

TABLE 11.5

Effect of Gamma Radiation on Polymers Containing Reinforcing Agents or Fillers

Polymer	Reinforcing Agent or Filler	Radiation Dose (kGy)	Measured Property	Reference
Polyether ether ketone	Carbon fiber	—	Molecular	58, 59
Polymethyl methacrylate	Hydroxyapatite	Up to 100	Melt flow index	54
Silicones	Silica	Up to 40 Mrad	Mechanical	57
Polybutylene terephthalate	Glass fiber	—	Thermal	60

Other polymers that have been the subject of beta irradiation studies include polyamide 64,65, polybutylene terephthalate [64], polyvinyl chloride [65, 66], and hydroxy propylmethyl/cellulose [67].

11.4　EFFECT OF ULTRAVIOLET IRRADIATION

Numerous tests are available for following photodegradation by ultraviolet irradiation. Infrared (IR) spectroscopy and measurements of change in the molecular weight (MW) and mechanical and electrical properties have been employed. Convenient tests have been found to be the measurement of flexural endurance for PE and change

TABLE 11.6
Ultraviolet Resistance of Polymers

Excellent Resistance	Very Good Resistance	Poor Resistance
PEEK	Cross-linked polyethylene	HDPE
Diallyl isophthalate	Polyisobutylene	LDPE
Ethylene-vinyl acetate	Styrene-ethylene-styrene	PP
PEI	PS	Polymethylpentene
Polyvinyl fluoride	Epoxies	Ethylene-propylene
PVDF	PET	Styrene-butadiene
Perfluoroalkoxy ethylene	Diallyl phthalate	Polyoxymethylene
Ethylene chlorotrifluoroethylene	Ethylene-tetrafluoroethylene	Styrene-maleic anhydride
Fluorinated ethylene-propylene	PC	Polyamide-imide
Silicones	Phenol formaldehyde	PI
	Polyamide 6	Polyurethane
	Polyamide 6,6	PMMA
	Polyamide 6,9	Urea-formaldehyde
	Polyamide 6,10	Styrene-acrylonitrile
	Polyamide 10	Acrylonitrile-butadiene-styrene
	Polyamide 2	Polyethersulfone
	Polyamide 12	
	PTFE	
	Chlorinated PVC	
	Unplasticized and plasticized PVC	
	PPS	
	Polysulfone	

PPS = polyethylene sulfide.

in MW and determination of hand brittleness for PP. Photodegradation is more intense at the surface of materials, so thin specimens have been used (25 mm) so that changes are more rapidly followed (although correlations have been established by means of fall-off in impact strength of thicker specimens). Change in the MW of PP is conveniently followed by measuring the change in solution viscosity of a 1% solution of polymer in decalin at 135°C. The critical MW below which a PP sample will exhibit brittleness on 2.5 mm thick samples has been found from tests on a wide range of samples to correlate with a limiting viscosity number (LVN) of 1.9.

Ultraviolet light assistance of polymers to weathering is an assessment of the effect of solar UV radiation at ambient temperatures. It is used as an overall assessment of changes in physical and mechanical properties and color. In Table 11.6, an *excellent* rating indicates excellent resistance to weathering and a *poor* rating indicates poor resistance to weathering.

It is seen that the most UV light-stable polymers are those based on fluorine and silicon. However, except for the most specialized of applications, most of these would be excluded based on cost. There remains a range of polymers that fall into the *very good* category of UV resistance, including some low-cost polymers such as polystyrene (PS), epoxies, and polycarbonate (PC). Polymers in the *poor* category would not generally be recommended for extended outdoor use unless they are protected by pigmentation or the use of UV-stabilizing additives.

11.4.1 PROTECTION AGAINST ULTRAVIOLET IRRADIATION

In contrast to ferrous metals, plastics do not suffer from the effect of corrosion. However, they are degraded by sunlight.

This results in deterioration of the mechanical and electrical properties of the polymer resistance to photooxidation by ultraviolet light varies. The following groups of plastics show decreasing light stability in the order of low-density polyethylene > high-density polyethylene > polypropylene > polystyrene.

Opaque pigments can have a valuable screening effect if incorporated into a plastic and can help to confine the photodegradation to the extreme surface layers, thus protecting the inner part of the plastic. The most effective of such pigments is carbon black. However, some care must be exercised in the choice of the grade of carbon black and its concentration to ensure maximum protection while avoiding side effects in other polymer characteristics.

Two percent by weight of a well-dispersed carbon black of fine particles in the range 15–40 μm should extend to outdoor life of most low-density polyethylene and all polypropylene grades to as much as 20 years in temperature climates and 76 years in tropical climates. Other pigments will have a much lower protective effect than carbon black.

Polypropylene is intrinsically less resistant to ultraviolet radiation. Although polypropylene shows powdering rather than grass surface grazing after sunlight exposure, minute surface cracks are visible under the microscope. Under certain conditions the microcracks can act as points of stress concentration. This can reduce the impact strength of the component.

Excellent protection is afforded by a range of carbon blacks on the weatherability of polypropylene. At a concentration of 2% of a carbon black in the 15–40 μm particle size range, polypropylene did not show a significant fall-off in molecular weight after 6000 sunshine hours exposure in a tropical climate. This would be almost 12,000 hours in temperate climates.

With either low-density polyethylene or polypropylene exposure to ultraviolet light under conditions of stress, after 1.5 years exposure was observed only with the unpigmented polymers; the pigmented materials are free of the surface-crazing marks exhibited by polymers not containing carbon black.

11.4.2 FORMULATION

Bramhiller et al. [68] have pointed out that outdoor degradation of isotactic polypropylene occurred after about 2520 h of exposure, after which oxygen species could be detected in the polymers.

An increase in density also occurred upon weathering, suggesting an increase in melt enthalpy.

Barony et al. [69] studied the effect of ultraviolet radiation on the tensile strength of syndiotactic polypropylene. Both tensile strength and strain values changed linearly with increased exposure to ultraviolet radiation and following an increase in the oxygen index of the polymer.

It has been observed [70] that the rate of degradation properties of polypropylene are affected by exposure to ultraviolet radiation.

REFERENCES

1. J.H. O'Donnell and D.J.J. Hill, in R.L. Clough and S.I.V. Shalaky (eds.), *Radiation Effect on Polymers*, ACS Symposium Series 475, ACS, Washington, DC, 1991, chap. 10.
2. A.G.A. Hafi, J.N. Hay, and D.J. Parker, *Journal of Polymer Science, Part B: Polymer Physics*, 2008, 462, 212.
3. S. Spielberger, K. Humer, K.K. Tsihagg, E.K. Weber, and H. Gerstevberg, *Cryoscopic Engineering*, 1996, 42, 1105.
4. M. Marinovic-Cincovic, E. Dzunuzovic, D. Balic, K. Popov-Rengal, and M. Pergal, *Polymer Degradation and Stability*, 2004, 86, 349.
5. C.L. Homrighausen, A.S. Mereness, E.J. Schutte, B. Williams, and N.E. Young, *High Performance Polymers*, 2007, 19, 382.
6. S.S. Cota, V. dos Vascone, M. Senne, L.L. Carvalho, D.B. Rezende, and R.F. Correa, *Brazilian Journal of Chemical Engineering*, 2007, 24, 251.
7. A.L. Babovich, Y. Unigovski, E.M. Gatman, E. Kolmakov, and S. Vyazovkin, *Journal of Applied Polymer Science*, 2007, 103, 3718.
8. I. Krupa and A.S. Luyt, *Polymer Degradation and Stability*, 2001, 71, 361.
9. L. Dao-Long, L. Peng-bo, and L. Ligin, *Polymer Materials Science and Engineering*, 2005, 21, 170.
10. A. Karam, Y. Sandez, N. Dominguez, G. Gonzalez, C. Alboro, and J. Reyes, *Molecular Crystals and Liquid Crystals*, 2004, 418, 189.
11. L. Lu, Y. Hu, Q. Kong, Z. Chen, and W. Fan, *Polymers for Advanced Techniques*, 2005, 16, 688.
12. J.C. Suarez, A. Elzubair, C.M.C. Bonelli, R.S. de Biagi, and E.B. Nano, *Journal of Polymer Engineering*, 2005, 25, 277.
13. D. Xiang J. Lin, D. Fan, and Z. Jin, *Journal of Materials Science: Materials Medicine*, 2007, 18, 213.
14. M. Kaci, H. Djidjelli, and T. Boukedami, *Polymer Bulletin*, 2008, 60, 387.
15. K.J. Kim, Y.H. Kim, J.H. Son, Y.N. Jo, J.S. Kim, and Y.J. Kim, *Journal of Power Sources*, 2010, 195, 6075.
16. M.M. Hassan, N.A. El-Kelesh, and A.M. Dessouki, *Polymer Composites*, 2008, 883.
17. Z. Stojanovic, Z. Kararevic-Popovic, S. Galovic, D. Milicevic, and E. Suijovrujie, *Polymer Degradation of Stability*, 2005, 87, 279
18. E. Nedkov and T. Dalreva, *European Polymer Journal*, 2004, 40, 2573.
19. M.F. Mina, S. Seema, R. Martin, M.J. Rahaman, R.B. Sarker, M.A. Enfur, and M.A.H. Bhuiyan, *Polymer Degradation and Stability*, 2009, 94, 183.
20. N. Tonato, M. Kaci, H. Ahouari, S. Bruzand, and Y. Groherg, *Macromolecular Materials and Engineering*, 2007, 292, 1271.
21. G. Akin-Okten, S. Tanrisinibilar, and T. Tincer, *Journal of Applied Polymer Science*, 2001, 81, 2670.
22. M.M. Hassan, N.A. El-Kelesh, and A.M. Dessouki, *Polymer Composites*, 2008, 29, 156.
23. V.I. Selikova, V.M. Neverov, E.A. Sinevich, V.S. Tikhomirov, and S.N. Chvalun, *Polymer Science, Series A*, 2005, 47, 103.
24. P.C. Kalst and A. Ramaswami, *Journal of Thermal Analysis and Calorimetry*, 2004, 78, 793.

25. J.C.M. Suarez, E.B. Mano, E.E. De Costa Monteiro, and M.I.B. Toveres, *Journal of Applied Polymer Science*, 2000, 85, 886.
26. S.M. Sayyah, M.A. El-Ahdal, Z.A. El-Shafiey, M. El-Sockary, and U.F. Kwondil, *Journal of Polymer Research*, 2000, 7, 97.
27. W.G. Hanno, M.A. Mekewi, and E.S.H. El-Mosallamy, *International Journal of Polymeric Materials*, 2003, 52, 471.
28. A.R. Souza, E.S. Aranjo, and M.S. Rabello, *Journal of Materials Science*, 2009, 44, 1035.
29. A.R. Souza, E.S. Araujo, A.L. Carvalho, M.S. Rabello, and J.R. White, *Polymer Degradation and Stability*, 2007, 92, 1465.
30. C.B. Bucknall and V.I.P. Soares, *Journal of Polymer Science, Part B: Polymer Physics*, 2004, 42, 2168.
31. W. Brostow, V.M. Castano, and G. Martinez-Barrera, *Polimery*, 2005, 50, 657.
32. W. Braston, V.M. Castano, J. Hortand, and G. Marinez-Barrera, *Int. Sci. Technol.*, 2004, 49, 9.
33. C. Albano, J. Reyes, M. Ichazo, J. Gonzalez, M. Hernandez, and M. Rodriguez, *Polymer Degradation and Stability*, 2002, 80, 251.
34. S.A. Nouh and S. Bahamman, *Journal of Applied Polymer Science*, 2010, 117, 3060.
35. C. Hon, J. Bian, H. Liu, and L. Dong, *Polymer International*, 2009, 58, 691.
36. M. Horia, N. El-Din, and M. Abdel-Wahob, *Polymer Composites*, 2008, 597.
37. M.M. Abou Zeid, S.T. Robie, A.A. Nada, A.M. Khalil, and R.H. Hilai, *Polymer Plastics Technology*, 2008, 47, 567.
38. T. Zhareseu, V. Meltzer, and R. Vilcu, *Polymer Degradation and Stability*, 1999, 64, 101.
39. E.B. Hermeda, V.I. Mega, O. Yashchuk, V. Fernandez, P. Eisenberg, and S.S. Miyazaki, *Microchemical Symposium*, 2008, 263, 102.
40. S. Green and K. Cartwright, *Proceedings of the Medical Polymers Conference*, Dublin, April 2–3, 2003, paper 5, p. 35.
41. V.V. Lavran'ev and G.V. Shiydnevski, *International Polymer Science and Technology*, 2010, 37, 143.
42. H. Simaoui, M. Mzabi, M. Guer Majo, Y. Matik, S. Agnel, A. Toureille, and F. Schul, *Polymer International*, 2007, 56, 325.
43. V.V. Lopatin, A.A. Askad Skii, V.G. Vasiliev, and E.A. Kurskaya, *Polymer Science, Series A*, 2004, 46, 751.
44. S.M. Powde and K. Deshmuskh, *Journal of Applied Polymer Science*, 2008, 110, 2569.
45. K. Gorner and S. Gogolewski, *Polymer Degradation and Stability*, 2003, 79, 465.
46. S. Vlad, *Materiale Plastice*, 2002, 39, 197.
47. C. Han, X. Ran, K. Zhang, Y. Zhung, and L. Dong, *Journal of Applied Polymer Science*, 2007, 103, 2676.
48. S.R. Allayarov, *Plasticheskie Massy*, 2006, 17.
49. A. Contkeros, E. Bucio, F. Leon, C.J. Booth, and D.E. Cassidy, *Polymer Preprints*, 2009, 50, 85.
50. G. Aarya, A.K. Siddhartha, A.K. Srivostava, A. Saha, and M.A. Wahab, *Nuclear Instruments and Methods in Physics Research*, 2009, 267, 3545.
51. C. Menchaca, L. Regon, A. Alvarez-Castillo, M. Apatiga, and V.M. Castono, *International Journal of Polymeric Material*, 2000, 48, 135.
52. W. Brostow, V.M. Castano, J. Horta, and G. Martinez-Barrera, *Polimery*, 2004, 49, 9.
53. T. Bremarer, D.T.T. Hill, J.H. O'Donnell, M.C.S. Pereri, and P.J. Powdes, *Journal of Polymer Science, Part A: Polymer Chemistry*, 1996, 34, 971.
54. R.G. Lehrle and C.S. Pattenden, *Polymer Degradation and Stability*, 1998, 61, 309.
55. R.S. Lehrle and C.S. Pattenden, *Polymer Degradation and Stability*, 1998, 62, 211.
56. N.N. Rupiasih and P.B. Vidyasagar, *Polymer Degradation and Stability*, 2008, 93, 1300.

57. A.S. Pasule and S.J. Clarson, *Journal of Inorganic and Organo Metallic Polymers*, 2008, 18, 207.
58. S. Green and K. Cartwright, *Proceedings of the Medical Polymers Conference*, Dublin, April 2–3, 2003, paper 5, p. 35.
59. C. Albano, A. Karam, G. Gonzalez, N. Domingnez, and Y. Sanchez, *Nuclear Crystal Liquid Crystals*, 2006, 448, 243
60. A Creanova Ltd., *European Polymer News*, 2000, 27, 71.
61. D. Krauz, V. Voigt, D. Auhl, and H. Munsted, *Journal of Applied Polymer Science*, 2003, 100, 2770.
62. R.M. Radwen, *Journal of Physics D*, 2007, 40, 374.
63. H.H. Babovich, U.N. Unigoiski, E.M. Gutman, and E. Kalmarkov, *Proceedings of 63rd SDE Annual Conference (ANTEC 2005)*, Boston, MA, May 15–25, 2005, p. 5.
64. Crean-Ova Ltd., *European Plastic News*, October 2000, citation 27, p. 71.
65. S. Thadani, H.W. Backham, and P. Desau, *Journal of Applied Polymer Science*, 1997, 65, 2013.
66. A.Y. Bamba, S.S. Kobyupaev, E.V. Kolupaev, E.V. Lebedev, I.M. Prisyazhnyuk, and A.M. Royal, *Journal of Polymer Science and Technology*, 2007, 34, T19.
67. D.T. Sangapp, D.S. Masindevkei, and D.M. Diveka, *Nuclear Instruments and Methods in Physics Research B*, 2008, 266, 3973.
68. L. Bramhiller, G. Constable, R. Gallo, F. Quasio, and F. Severini, *Polymer*, 2003, 44, 106.
69. J. Barony, E. Folder, T. Czigony, and J. Kanger-Koksis, *Journal of Applied Polymer Science*, 2004, 91, 3462.
70. J. Mayumdan, F. Cser, M.C. Jollandi, and R.A. Shanks, *Journal of Thermal Analysis and Calorimetry*, 2004, 78, 849.

12 Unreinforced and Reinforced Fire Retardant Polymers

In many applications of engineering polymers, there is a requirement for fire retardant properties. This will occur in appliances of polymers used in situations were excessive heat or long periods of time may occur or the plastic part is used in a region where there is risk of ignition and fire. This might be particularly so in electrical applications of reinforced polymers.

12.1 FLAMMABILITY CHARACTERISTICS

As would be expected, the incorporation of a fire retardant additive improves all fire testing parameters to the good or very good category. Higher additions of lined fillers may, in some cases, improve fire retardancy characteristics over those obtained for a virgin fire retardant polymer. This occurs in the case of fire retardant polyphenylene oxide, where, as shown in Table 12.1, the addition of 30% glass fiber to the virgin polymer improves flame spread from the good to the very good category.

In the case of some non–fire retardant grades of polymers fire retardancy characteristics might improve when a filler is incorporated into the formulation. Thus, in the case of epoxy resins, the incorporation of minerals, glass fiber, silica, or graphite improves flame spread, flammability, and LOI from the poor category in the virgin polymer to the good category.

Such improvement cannot, however, be taken for granted. No improvement in fire retardancy characteristics was observed when a wide range of filler/reinforcing agents were incorporated into non–fire retardant polycarbonate, polybutylene terephthalate, polyethylene terephthalate, polyamide 6,6, or styrene acrylonitrile copolymer.

In some instances, the incorporation of a fire retardant can affect mechanical, electrical, and thermal properties of a polymer.

Three factors should be considered before any decisions are made on the selection of a fire retardant polymer for a particular application.

The incorporation of a reinforcing agent such as glass fibers into a non–fire retardant or a fire retardant formulation might also have an effect on polymer properties, as discussed below.

12.2 EFFECT ON PHYSICAL PROPERTIES

12.2.1 MECHANICAL PROPERTIES

Available information on mechanical properties of various polymers is tabulated in Table 12.2.

TABLE 12.1

Comparison of Fire Retardancy Properties of Non–Fire Retardant and Fire Retardant Grades of Polymers Containing Various Reinforcing Agents

Polymer	Filler	Non–Fire Retardant Grades			Fire Retardant Grades		
		Flame Spread	Flammability	LOI	Flame Spread	Flammability	LOI
Epoxies	Virgin	Poor	Poor	Poor			
	Mineral and glass fiber	Good	Good	Good	Very good	Very good	Very good
	Minerals	Good	Good	Good			
	Glass fiber	Good	Good	Good			
	Silica	Good	Poor	Good			
	Carbon fiber	Good	Good	Good			
Polyphenylene oxide	Virgin	Poor	Poor	Poor	Good	Good	Good
	30% glass fiber	Poor	Poor	Good	Good	Good	Good
	30% glass fiber	Poor	Poor	Good	Very good	Good	Good
Polycarbonate	Virgin	Good	Poor	Poor	Very good	Very good	Very good
	15% PTFE	Good	Poor	Poor			
	30% carbon	Poor	Poor	Poor			
	20% glass fiber	Good	Poor	Poor			
Polyester	Virgin	Good	Good	Poor	Very good	Very good	Very good
Diallyl phthalate	Virgin	Poor	Good	Good	Very good	Very good	Good

Polybutylene terephthalate	Virgin	Poor	Poor	Poor	Very good	Good	Good
	30% glass fibee	Poor	Poor	Poor	Very good	Good	Good
Polyethylene terephthalate	Virgin	Poor	Poor	Poor	Very good	Good	Good
	30% glass fiber	Poor	Poor	Poor	Very good	Good	Good
	35% glass fiber	Poor	Poor	Poor			
	Mineral	Poor	Poor	Poor			
	45% glass fiber	Poor	Poor	Poor			
	15% glass fiber	Poor	Very poor	Very poor			
	35% MICA	Poor	Poor	Poor			
Polyamide 6,6	Virgin	Poor	Poor	Poor	Very good	Good	Good
	10% carbon	Poor	Poor	Poor			
	30% carbon	Poor	Poor	Poor			
	20% PTFE	Poor	Poor	Poor			
	40% mineral	Poor	Poor	Poor			
	60% glass fiber	Poor	Poor	Poor			
Styrene-acrylonitrile	Virgin	Poor	Very poor	Poor	Good	Good	Good
	30% glass fiber	Very poor	Very poor	Poor			
Acrylonitrile butadiene styrene	Virgin	Poor	Poor	Poor	Very good	Good	Good

TABLE 12.2

Comparison of Mechanical Properties of Filled and Unfilled Non–Fire Retardant and Fire Retardant Goods

Polymer	Filler Fiber	Non–Fire Retardant Grades					Fire Retardant Grades				
		Tensile Strength, MPa	Flexural Modulus, GPa	Degradation at Break, %	Strain at Yield, %	Notched Izod Impact Strength	Tensile Strength, MPa	Flexural Modulus, GPa	Elongation at Break, %	Strain at Yield, %	Notched Izod Impact Strength, kJ/m
Polyimide 6,6	Nil	Good	Very poor	Very poor	Poor	0.11	Poor	Very poor	Good	Very poor	—
Styrene-acrylonitrile	Nil	Good	Good	Poor	Poor	0.02	Good	Poor	Poor	Poor	—
Polybutylene terephthalate	30% glass fiber	Very good	Very good	Poor	Very poor	—	—	Poor	—	—	—
Polybutylene oxide	Nil	Good	Poor	Poor	Poor	0.16	Good	Poor	Good	Good	—
	10% glass fiber	Good	Poor	Good	Very poor	—	Good	Good	Good	Very poor	—
	30% glass	Very good	Very good	Poor	Very poor	—	Very good	Very good	Poor	Very poor	—
Polypropylene	Nil	Poor	Very poor	Very good	Good	0.05	Very poor	Very poor	Good	Good	—
Polyesters	Nil	Poor	Good	Poor	Very poor	Greater than 1.06	Poor	Good	Poor	Very poor	—

The incorporation of 10%–30% of glass fiber into the formulation of either non–fire retardant or fire retardant grades produces improvements in tensile strength and flexural modulus and a deterioration of percent elongation at break and, in the case of the fire retardant grade, percent strain at yield.

12.2.2 ELECTRICAL PROPERTIES

The incorporation of a fire retardant additive, in the case of some polymers, produces a deterioration in surface arc resistance and tracking resistance. This has been observed in the case (Table 12.3) of diayllyl phthalate, polyamide 6,6 and polybutylene terephthalate. A slight deterioration in dissipation factor was also observed in the cases of polybutylene terephthalate, polyphenylene oxide, and polypropylene.

The incorporation of 10%–30% glass fiber into the non–fire retardant and fire retardant polyphenylene oxide formulations causes a decrease in dielectric strength.

12.2.3 THERMAL PROPERTIES

The only effects on thermal properties of the incorporation of a fire retardant additive occur in the case of high-impact polystyrene, where, as shown in Table 12.4, the incorporation of a fire retardant leads to a decrease in the expansion coefficient, and in the case of polyesters, where the incorporation of a fire retardant produces a small improvement in heat distortion temperature.

TABLE 12.3
Comparison of Electrical Properties of Filled and Unfilled Non–Fire Retardant and Fire Retardant Grades

Polymer	Filler	Non–Fire Retardant Grades						Fire Retardant Grades					
		Volume Resistivity, ohm/cm	Dielectric Strength, mV/m	Dielectric Constant, 1 kHz	Dissipation Factor, 1 mHz	Surface Arc Resistance, ohm	Tracking Resistance, ohm	Volume Resistivity, ohm/cm	Dielectric Strength, mV/m	Dielectric Constant, 1 kHz	Dissipation Factor, 1 mHz	Surface Arc Resistance, ohm	Tracking Resistance, ohm
Polycarbonate	Nil	Very good	Good	Very good	Poor	Very poor	Poor	Very good	Good	Very good	Poor	Very poor	Very poor
Diallyl phthalate	Nil	Good	Poor	Good	Poor	Excellent	Excellent	Good	Poor	Good	Poor	Good	Very good
Polyamide 6,6	Nil	Good	Poor	Good	Poor	Excellent	Excellent	Good	Poor	Good	Poor	Good	Very good
Styrene-acrylonitrile	Nil	Good	Very good Very good Poor	Good	Poor	Good	Good	Very good Poor	Good	Poor			
Polybutylene terephthalate	Nil	Good	Good	Very Poor	Poor	Very good	Good	Good	Good	Very good Very poor Good	Very poor		
Epoxy resins	Mineral	Good	Poor	Poor	Very good	Very good	Good	Good	Very poor	Good	Very good	Very good	Poor
Polyphenylene oxide	Nil	Very good	Good	Very good	Very Good	Very poor	Poor	Very good	Good	Very good	Good	Very poor	Poor
	10% glass fiber	Good	Poor	Very good	Very good	Poor	Poor	Good	Good	Very good	Good	Very good	Very poor
	30% glass fiber	Very good	Poor	Very good	Very good	Good	Very poor	Very good	Very poor	Very good	Good	Very poor	Very poor
Polypropylene	Nil	Good	Very good	Excellent	Excellent	Excellent	Excellent	Good	Good	Excellent	Good	Very good	Very good
Polyesters	Nil	Good	Very poor	Poor	Poor	Excellent	Excellent	Good	Very poor	Poor	Poor	Excellent	Excellent

TABLE 12.4

Comparison of the Thermal Properties of Filled and Unfilled Non–Fire Retardant and Fire Retardant Grades

Polymer	Filler	Non-Fire Retardant Grades				Fire Retardant Grades			
		Expansion Coefficient, Mm/°C x 10	Heat Distortion Temperature at 1.8 MPa °C	Brittle Temperature, %	Mold Shrinkage, %	Expansion Coefficient, Mm/°C x 10	Heat Distortion Temperature at 1.8 MPa °C	Brittle Temperature, %	Mold Shrinkage, %
Polycarbonate	Nil	Good	Good	Good	Good	Good	Good	Good	Good
Styene-acrylonitrile	Nil	Very good	Good	Very poor	Good	Good	Poor	Very poor	Good
Polybutylene terephthalate	Nil	Poor	Poor	Poor	Very poor	Poor	Poor	Very poor	Very poor
	30% glass fiber	Good	Very good	Very poor	Good	Good	Very good	Very poor	Good
Epoxy resins	Glass fiber	Very good	Very good	Very poor	Very good	Very good	Very good	Very poor	Very good
Polypropylene	Nil	Poor	Poor	Very poor	Poor	Poor	Poor	Very poor	Poor
High-impact polystyrene	Nil	Good	Poor	Poor	Good	Poor	Poor	Poor	Good
Polyesters	Nil	Very good	Good	Poor	Good	Very good	Excellent	Poor	Very good
Polyamide 6,6	Nil	Good	Poor	Very Good	Poor	Good	Poor	Good	Poor

Index